Earth
FROM SPACE

DATE			

Earth
FROM SPACE

SMITHSONIAN NATIONAL AIR AND SPACE MUSEUM

Andrew K. Johnston

FIREFLY BOOKS

TITLE PAGE
Himalayan Mountains

The Himalayan Mountains separate India from the rest of the Asian continent. They were formed by the collision of two massive plates on the Earth's crust which buckled and folded to form the world's largest mountain range. India is still being pushed slowly northward into Asia. This image was produced by combining data from different sources. Topographic data were used to increase the clarity of mountains and other large features. The image data were collected by the MODIS sensor on the Terra satellite in June–September 2001.

OPPOSITE CONTENTS PAGE
New Lakes in the Desert

This image, acquired by the Landsat 7 satellite on January 4, 2003, shows new lakes being created in the Western Desert of Egypt. Water resources are carefully controlled because little or no rain falls in this area. The Aswan High Dam was completed on the Nile River in 1970, which created Lake Nasser about 90 miles (150 km) east of the area shown in this image. Beginning in 1997 new canals were dug through the desert to divert water from Lake Nasser as part of development plans. This image, which uses infrared light, shows features such as sand dunes being submerged as the water rises.

Earth From Space exhibit

This traveling Smithsonian exhibit, based on the *Earth From Space* book, began its national tour in 2006. See the web site to learn about the images in this book and for more information about the exhibit: **www.earthfromspace.si.edu**

A FIREFLY BOOK

Published by Firefly Books Ltd. 2007

Copyright © 2007 Andrew K. Johnston

First printing

Publisher Cataloging-in-Publication Data (U.S.)
Johnston, Andrew K. (Andrew Kenneth), 1969-
 Earth from space : Smithsonian National Air and Space Museum / Andrew K. Johnston.
2nd rev. and updated.
[272] p. : col. ill., col. maps ; cm.
Includes bibliographical references and index.
Summary: Three hundred satellite photographs illustrate how satellite imaging works. Subjects covered include aeronautics, history and ecology.
ISBN-13: 978-1-55407-291-0 (pbk.)
ISBN-10: 1-55407-291-3 (pbk.)
1. Earth — Photographs from space. 2. Space photography. 3. National Air and Space Museum.
I. Title.
910/.22/2 dc22 QB637.J55 2007

Library and Archives Canada Cataloguing in Publication
Johnston, Andrew K. (Andrew Kenneth), 1969-
 Earth from space : Smithsonian National Air and Space Museum / Andrew K.
Johnston. — 2nd ed., rev. and updated
Includes bibliographical references and index.
ISBN-13: 978-1-55407-291-0
ISBN-10: 1-55407-291-3
1. Earth—Photographs from space. 2. Space photography. I. National Air and
Space Museum II. Title.
QB637.J63 2007 910.22'2 C2007-900276-5

Published in the United States by
Firefly Books (U.S.) Inc.
P.O. Box 1338, Ellicott Station
Buffalo, New York 14205

Published in Canada by
Firefly Books Ltd.
66 Leek Crescent
Richmond Hill, Ontario L4B 1H1

Cover and book design by Kevin Osborn / Research & Design Ltd., Arlington, Virginia

Printed in China

The publisher gratefully acknowledges the financial support for our publishing program by the Government of Canada through the Book Publishing Industry Development Program.

Acknowledgments

Earth from Space was made possible by many people who contributed their hard work, time and talents. The research, writing and image compilation were performed at the Center for Earth and Planetary Studies (CEPS), National Air and Space Museum. Many thanks go to all the staff at CEPS, who provided help in countless ways. Kenneth Brown organized image collections, assisted with the production of maps and graphics, and performed background research. Thanks to Sharon Wilson for her assistance.

Many thanks go to Trish Graboske, head of publications at the National Air and Space Museum, for her help and insights. This book would not have been possible without the expert guidance of Chuck Hyman. *Earth from Space* was an extremely complex project, requiring the manipulation of hundreds of digital image files. Designers Kevin Osborn and Linda Gustafson did an excellent job handling the massive loads of data. Thanks to editor Dan Liebman for his countless improvements to the text. Many thanks to Janet Johnston for reviewing the text for the second edition.

Many thanks also to the reviewers, three Smithsonian colleagues who provided their expertise in a wide range of fields: Michael Harrison, National Museum of American History, James Zimbelman, National Air and Space Museum, and Geoffrey Parker, Smithsonian Environmental Research Center. Anne-Louise Marquis, Raywat Deonandan and Sneh Aurora reviewed text on specific subjects. The insights they provided made this a much better book.

Earth from Space would not have been possible without the many people who provided the spectacular images in these pages. The images came from a wide range of commercial firms, universities and government agencies. The following people, listed in random order, helped tremendously by acquiring images and allowing their use for this publication: Chuck Herring, DigitalGlobe; Mark Brender, GeoEye; Val Webb, GeoEye; Clark Nelson, SPOT Image; Byron Loubert, MDA Federal; Karen Gold Anisfeld, ImageSat International; Leopold Romeijn, Satellite Imaging Corporation; Lisa Andrews, ORBIMAGE; Vitaly Ippolitov, R&D Center Scanex; Ken Jezek, Ohio State University; William Stefanov, Arizona State University; Julie Robinson, NASA Johnson Space Center; Michael Abrams, NASA/JPL; David Diner, NASA/JPL; Dennis Chesters, NASA Goddard Space Flight Center; Norman Kuring, NASA Goddard Space Flight Center; Fritz Hasler, NASA Goddard Space Flight Center; Peter Reinartz, DLR; Ron Beck, USGS/EROS Data Center.

Extra thanks to David Diner, Chuck Herring, Ron Beck, Karen Gold Anisfeld, Dennis Chesters, Michael Abrams, Vitaly Ippolitov, William Stefanov, Val Webb and Byron Loubert, all of whom provided image data especially for this book.

Many thanks to David Leverington and Sarah Andre for providing frequent advice to the colorblind author. Thanks also to Kevin Williams, who provided spell-checking assistance when it mattered most.

Contents

The Global View

The Earth, our home in space, is a varied and dynamic place. Since the beginning of human history we have sought a better understanding of the world around us. With the advent of the aerospace age we can look back and appreciate the diversity and beauty of the Earth in a way not possible until the 20th century.

These are exciting times to be observing the Earth from space. Since the mid-1990s a new generation of satellites — with powerful capabilities to collect massive amounts of data — has been launched.

This book provides an understanding of how these amazing satellites work and how their images are increasingly important in many aspects of our lives. It presents many new images that were collected to serve scientific or technical needs, but that can often be appreciated simply for their beauty. The sensors that acquired the images see things in different and more powerful ways than our eyes do. The imaging satellites now in orbit provide perspectives that give us a whole new appreciation for the planet on which we live.

Remote Sensing

Remote sensing is a simple term that can be applied to any technology that views something from a distance. In the case of understanding the Earth, remote sensing means observing the Earth from above, either from an airplane or from a satellite in space.

Two technologies, photography and aeronautics, converged in the late 1800s to change the way we look at our world. Initial attempts to obtain a remote view of the Earth involved kites and balloons. As technology progressed, airplanes were seen as a way to obtain views of the Earth not possible from the ground. Crews on airplanes could carry simple cameras and take photos. As satellites were launched

Earth

This view (opposite page) shows Earth as it appears from a satellite orbiting high over the Indian Ocean. Darker colors on the land surfaces are areas covered by forests and other vegetation; the lighter brown indicates arid areas. Much more detail is visible here than in our first global views, which were obtained in the late 1960s by Apollo astronauts returning from the Moon. Data from several different satellite sensors, especially the MODIS instrument on the Terra satellite in mid-2001, were combined to produce this image.

Clarity of the Atmosphere

This global map shows aerosol optical depth, which is a measure of the density of aerosol particles in the atmosphere. These particles include smoke, dust, volcanic ash and ocean salts. Yellow areas have high aerosol concentrations; light blue shows low aerosol concentrations. Aerosols tend to prevent sunlight from reaching the Earth's surface by reflecting or absorbing light. They are important in helping us understand how energy is dissipated in the atmosphere and how global temperatures are determined. Satellite sensors can detect the amount of light absorbed by aerosols and the amount reflected back into space. This image was made from data collected by the MODIS sensor on the Terra satellite.

into space, the possibilities of remote sensing from space were widely recognized. Some of the first satellites carried cameras for observing clouds in the atmosphere. Later satellites were designed to observe the land surface. Current technology allows satellites to carry a variety of instruments, including cameras and digital sensors. There are dozens of remote sensing satellites in orbit with a wide range of capabilities to help us understand the surface of the Earth.

There are many different ways to collect images of the Earth from space. Reflected sunlight can be used to discern physical characteristics of the surface. Thermal-infrared radiation can be detected to determine temperature. Satellites equipped with radar can see through clouds and at night. Images taken at different times can be used to detect changes. Satellite images can be combined to provide three-dimensional views of the Earth's surface.

There are a great many applications for remote sensing data. Planners and cartographers use satellite images to map the Earth. Military commanders use the images to plan missions. Biologists can map density and diversity over huge areas. Meteorologists use satellites to observe and predict the movement of weather patterns. Geological differences can be mapped from orbit as well. New satellites continue to expand the possibilities for these images.

Geography

Geography is the study of people, objects and processes on the Earth in their spatial context, and this book is a geographical exploration of the world we live in. The use of satellite images allows us to understand where features in the atmosphere, oceans and land surface are located and how they interact.

We can observe the Earth from a great distance to obtain a global view, or we can look closely to see small details. Held at arm's length, a simple globe gives an observer a view of the entire planet. It is possible to see how the continents are arranged, but smaller features are not visible. Moving the globe closer to our eyes brings rivers and mountain ranges into view, but now it is no longer possible to see the entire planet. To see details such as city streets, other tools, such as city maps, are necessary.

Geographers use the term *scale* to describe the relationship between the Earth's surface and its representation on a map or globe. Satellite images come in many different scales. With the diversity of satellites now in orbit, we can observe the entire globe or zoom in on small areas. There is an appropriate scale for each remote sensing application.

Some satellites are in high orbits, thousands of miles above the Earth. Designed to observe an entire hemisphere, they are useful for understanding processes, such as the changing weather, that operate throughout the entire globe. However, they cannot discern small details.

Other satellites are in lower orbits, a few hundred miles high. From their vantage point it is possible to observe large areas such as entire countries or large metropolitan areas. They can also discern smaller things on the Earth's surface, such as very large buildings or agricultural fields, making them useful for mapping growing cities or regional environmental changes.

Still other satellites are in slightly lower orbits, with advanced cameras capable of discerning small details. These satellites are useful for planning and military applications since objects the size of an automobile are visible in their images. If researchers are locating small buildings or planning where to build a street, they would use one of these satellites. Because of the amount of visible detail, the images from these satellites can cover only small areas, such as a city's downtown.

As a result of our growing knowledge of how to use satellite images, information from different scales can be combined to provide a better understanding of the Earth. The technology of building and launching spacecraft has changed dramatically since the first satellite was launched in 1956. The technology of processing data on the ground has also advanced. New computer-processing techniques allow scientists to create global-scale maps by stitching together many images taken from low-orbit satellites. Although these images may be taken at different times, through advanced processing techniques they can be combined to cover the entire Earth in one large image.

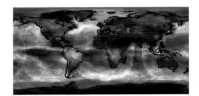

Global Cloud Cover

This global map shows the proportion of the sky covered by clouds. The lighter color represents heavy cloud cover. The amount of cloud cover is important because clouds reflect sunlight back into space and trap heat within the atmosphere. These global maps were produced from data collected by geostationary weather satellites.

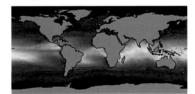

Temperature of the Seas

Satellites can take the temperature of the Earth's oceans. This information is used to follow changes in the global environment, to predict the strength of weather systems and to track ocean currents. Ocean temperature influences the growth of phytoplankton in the oceans and precipitation patterns across continents. Data from the MODIS sensor on the Terra satellite were used to make this image, where brighter colors represent higher temperatures.

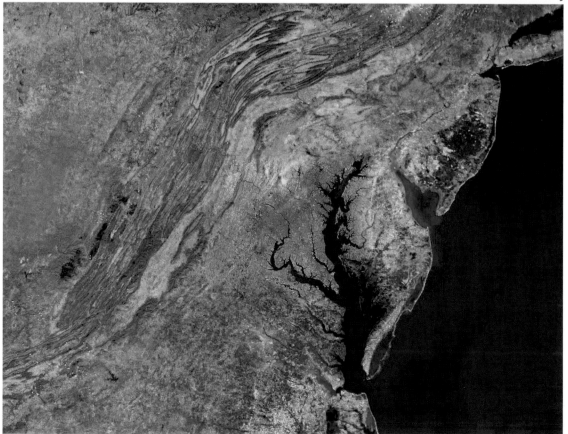

▶ 1

▲ 2   ▼ 3

Global-to-Local Scales

Satellites view the Earth at different scales. These images were produced by combining data from several satellites. The entire Earth is visible from weather satellites in geostationary orbit. The first image (1) shows this type of view, far above North America.

The eastern United States and Canada are visible in the second image (2). This is an appropriate scale for observing environmental changes that affect large areas. Only the eastern United States is visible in the third image (3). Satellites such as Aqua and Terra observe the Earth at this scale.

The next image (4) zooms into the Washington DC metropolitan area. This is the detail visible to satellites such as the Landsat and SPOT.

At the scale of the next image (5), individual buildings are easily visible. Images like this one, used for city-planning proposes, are provided by commercial high-resolution satellites such as IKONOS and QuickBird.

The high-resolution view (6) zooms into the U.S. Capitol Building. This is the highest resolution available to nonmilitary satellites.

▲ 4   ▼ 5   ▼ 6

Sand in the Desert

ARABIAN PENINSULA

One of the world's driest regions — the Arabian Peninsula between the Red Sea and the Persian Gulf — lies in the center of this image. The central part of Saudi Arabia is dominated by this kind of landscape. Bright yellow indicates huge amounts of loose sand. Many images from the Landsat satellite were combined to produce this image.

Tools of the Trade

For most of the 20th century people devised methods to obtain views from above. Balloons, airplanes and rockets were used to take pictures of the Earth's surface. Space exploration in the 1960s and 1970s allowed us to go high enough into space to see the entire Earth in one view, changing forever the way we understand our planet. Today, there are dozens of satellites in orbit designed to acquire images of the Earth's surface.

Early Beginnings

The first photographs from above the Earth's surface were taken from a balloon. Some of the earliest were published in 1858, showing suburbs of Paris. In 1895 an 18-foot-high kite was flown over an army barracks in New York. It carried a camera that used a timer to snap photos from an altitude of 600 feet (180 m). In 1903 a German patent was obtained for a camera small enough to be strapped to a pigeon. The camera was later used to obtain good-quality photos above Kronberg, Germany.

In 1860 the first successful photographs from the air in North America were taken from a balloon high above Boston. Remote sensing experiments with balloons became more widespread during the American Civil War. In 1861 Thaddeus Lowe flew balloons above battlefields to track the movements of rebel armies and target artillery. That year, during a balloon flight over Washington DC, Lowe sent a telegraph message from an altitude of 500 feet (150 m).

Balloons continued to be used in modern times, and in 1956 the United States, as part of a reconnaissance program called Moby Dick, released high-altitude balloons equipped with cameras. These balloons were to float over the Soviet Union, after which their cameras could be recovered. Of more than 500 such flights, however, only 44 were successful.

Caribbean Harbor
ST. JOHN'S, ANTIGUA
This satellite image (opposite page) shows St. John's harbor with a large cruise ship docked next to two sailing vessels. St. John's is the capital of the Caribbean island nation of Antigua and Barbuda. The commercial heart of the city stretches east from the harbor.

Although Columbus named the island in 1493, he never landed there. In the 1600s sugarcane cultivation became widespread on the island. Today, tourism is a main pillar of Antigua's economy. This high-resolution image was made by the IKONOS satellite.

Visible and Infrared Light

These two images show the Okavango River flowing into northern Botswana. As it flows past vegetated sand dunes, it widens into a broad inland delta. The waters of the Okavango Delta eventually dry up in Botswana's arid interior.

These two images, collected by the Landsat 7 satellite at exactly the same time, illustrate how different wavelengths of light can be used to make different images. The top image shows the area as it appears in visible light. During processing of the bottom image, reflected near-infrared light was assigned the color red. The delta's vegetation reflects strongly in those wavelengths.

Into the Air

In 1909 Wilbur Wright piloted the first airplane to carry a movie camera. During World War I, aircraft were used for reconnaissance. Working out of open biplanes, photographers had to operate under freezing and windy conditions. Later, in the 1920s, many types of cameras — including ones with electric shutters and those with multiple apertures that provided distortion-free views of wide areas — were designed to be used from aircraft.

Aerial reconnaissance became an integral part of military planning in World War II. During the Cold War airplanes equipped with cameras continued to provide important defense information. In the 1950s U-2 airplanes began to fly missions for U.S. intelligence agencies. These aircraft could use their 80-foot (25 m) wingspan to fly at high altitudes, where they were invulnerable until longer-range missiles came into use. Later, the SR-71 Blackbird, which could fly much faster, took over this intelligence-gathering duty.

Rockets into Space

Rocket pioneer Robert Goddard experimented with photography in the 1920s, and in 1925 a Goddard rocket took the first successful night photograph from above. Residents of Rochester, New York, were stunned when an 80 pound (36 kg) flash bomb exploded in the sky, providing the light for a photograph.

The first artificial satellites were launched in 1956, ushering in a new age of space exploration. Most orbital remote sensing was performed for military reconnaissance. During the Cold War both the United States and the Soviet Union launched hundreds of satellites equipped with cameras. Although they functioned much like conventional cameras, the devices on satellites were much larger and were designed to detect objects as small as parked aircraft.

In the late 1950s the U.S. Air Force began the Corona Project, an orbital reconnaissance program with the cover name Discoverer. Large cameras launched into space collected photographs over the Soviet Union and other important areas. The exposed film was then returned in a capsule to Earth. In August 1960 the first capsule was successfully recovered, making it the first object to return to Earth from orbit. The first Corona mission to return film collected more imagery of Soviet territory than did all the U-2 flights combined. Over the next decade more than 100 Corona flights returned about 800,000 photographs. The program ended in 1972, when digital technologies surpassed the capabilities of camera systems.

Modern Military Applications

Today's military services use remote sensing data to define objectives and provide air and ground forces with the information they need. Remote sensing has been used in coordination with other technologies to reshape modern warfare. Air power before 2000 used mostly unguided bombs, which were so inaccurate that fewer than one percent scored direct hits. During military action in Afghanistan in 2002 and Iraq in 2003, bombs and missiles used by U.S. forces hit their targets about 90 percent of the time.

There are many misconceptions about the capabilities of the military in orbit. Hollywood films often show moving images from space that resemble scenes in home movies. Sometimes the hero of the movie is able to target satellites immediately and search for evildoers from the comfort of home. However, real satellites in orbit pass over a particular point on the Earth's surface every couple of weeks and cannot be moved over any site at any time.

Immediate remote sensing data and moving images can be provided by unpiloted aircraft. The use of drone aircraft equipped with cameras and flying in battlefield conditions is now routine. These aircraft can provide real-time imagery in many wavelengths. They can even be equipped with weapons to attack targets that are found.

More money and people are involved in military remote sensing than in civilian applications. United States intelligence agencies and military services use more satellite images than do all other users combined. Although the specific characteristics of military remote sensing satellites are kept secret, some basic outlines of their designs are widely known. The most advanced military reconnaissance satellites provide images that make it possible to identify cars and perhaps even individual people getting into them.

Although there are important differences, nonmilitary satellites provide similar data for civilian applications. Biologists use them to understand life on Earth. Geologists are able to observe differences in the Earth's surface, allowing them to better understand geological history. Miners use satellite images to locate promising sites in the search for petroleum or valuable mineral deposits. Meteorologists use satellites to track weather patterns. Urban planners use them to map cities and to plan for future development. Firefighters use satellite images to identify where forests are burning and, afterward, to identify the burn scars.

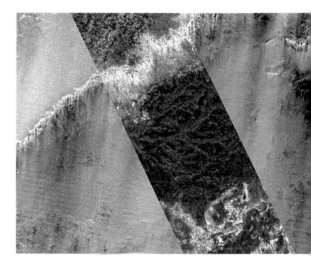

Radar Seeing beneath the Sands
The central strip of radar data from the SIR-C Space Shuttle mission reveals many dry drainage channels that lie beneath the sands of Egypt's Western Desert near the Bir Kiseiba oasis.

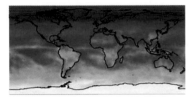

Shortwave Radiation

Sunlight has a wide range of wavelengths. Shortwave radiation includes all the light we can see with our eyes as well as much of the infrared spectrum. About 30 percent of this radiation is reflected back into space by the atmosphere, the oceans and the land surface. Oceans reflect more shortwave radiation than land surfaces. This image shows the amount of shortwave radiation reflected back into space. Bright colors indicate high sunlight reflection, and blue colors indicate lower reflected sunlight. These global images were made six months apart by the CERES sensor on the Terra satellite.

Cameras and Digital Sensors

The Space Shuttle has taken very large cameras on some of its missions. In 1983 the ninth Space Shuttle flight carried the Metric Camera, which collected about 1,000 photographs. The Large Format Camera, flown on the Space Shuttle in October 1984, weighed 1,000 pounds (450 kg) on the Earth's surface.

Astronauts on the first Mercury missions in the early 1960s carried 35mm cameras. Later missions took 70mm cameras on board. Astronauts on the Space Shuttle and the International Space Station still use this type of large-format camera. After 2000 many photographs from the Space Station, including some of the images in this book, were taken with digital cameras.

Beginning with the first Landsat satellite in 1972, most remote sensing satellites have carried digital sensors. These machines observe the Earth's surface by digitally encoding brightness values.

Many Sensors in Orbit

The 1972 launch of the first Landsat satellite marked the start of a new era of satellite remote sensing. Originally called the Earth Resources Technology Satellite, it was equipped with the Multi-Spectral Scanner (MSS), which had the ability to sense four wavelengths — two visible and two infrared. Two more Landsat satellites followed in 1975 and 1978. In 1982 and 1984, Landsat 4 and 5 were launched. They were equipped with the Thematic Mapper (TM) sensor, which had the ability to detect seven different wavelengths, including thermal infrared light. The next satellite in the series, Landsat 6, did not achieve orbit.

Landsat 7 was launched in 1999, but it suffered a partial failure of its Enhanced Thematic Mapper plus (ETM+) in 2003. Since Landsat 4, all satellites in the series have been placed into similar orbits 438 miles (705 km) in space. These satellites revolve around the Earth in a north-south direction, crossing over each point on the Earth's surface every 16 days.

Dozens of other satellites also carry sensors for observing the atmosphere and the Earth's surface. The GOES satellites, operated by the National Oceanographic and Atmospheric Administration, observe weather patterns from geostationary orbits. The satellites in the SPOT series were launched by the European Space Agency, which also operates the Envisat satellite. Terra and Aqua are large NASA satellites that carry several environmental sensors. Some satellites have used radar to observe the Earth's environment, including the RADARSAT satellites, developed by the Canadian Space Agency.

Most remote sensing satellites are operated by government agencies, but since the late 1990s a new generation of satellites has been placed into orbit by private companies. They include the OrbView series and the IKONOS satellite, operated by GeoEye; the QuickBird Satellite, owned by DigitalGlobe; and the EROS satellites launched by ImageSat International. The OrbView, IKONOS, QuickBird and EROS satellites provide very detailed imagery of small areas. A complete list of remote sensing satellites appears in the back of this book.

How It Works

Most remote sensing satellites observe the Earth using reflected sunlight. The Sun, a middle-aged star at the center of our solar system, provides all the light. This light travels through space to the Earth's surface and is reflected back into space, where it is detected by satellites.

The Sun's light is made up of a wide range of wavelengths, most of which we cannot see with our eyes. The full range of wavelengths is called the electromagnetic spectrum. The small range of light that we can see with our eyes, called visible light, accounts for about 40 percent of the energy in solar radiation. Visible light is used by satellites for observing and mapping changes on the Earth's surface. Orbiting sensors can detect many other wavelengths, including: near-infrared light, which is important for understanding the location and health of vegetation; other infrared wavelengths useful for examining geological features; and thermal infrared, used to detect heat. Some ultraviolet wavelengths are used to measure characteristics of the Earth's atmosphere.

The digital sensors on satellites detect the amount of light reaching the satellite and record that amount as a long string of numbers. Higher numbers indicate brighter values. These long streams of data are then radioed to receiving stations. The images made from the data contain millions of pixels. Each pixel is represented by a number value for each wavelength of light.

Multispectral sensors on orbiting satellites are designed to detect light from many different wavelengths (or channels) simultaneously. The satellite sends back several different black-and-white images, one for each channel. Computer programs are then used to produce a full-color image.

The capabilities of multispectral sensors vary with each satellite. The ETM+ sensor on Landsat 7 can detect eight wavelengths. The MODIS sensor on the Terra satellite can detect 36. Another class of instruments, called hyperspectral sensors, can detect hundreds of wavelengths. The range of wavelengths detected is called the spectral resolution of the sensor.

Radar sensors use the microwave portion of the electromagnetic spectrum.

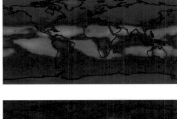
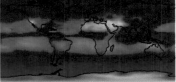

Longwave Radiation
Solar radiation that is absorbed by the Earth is stored as thermal energy, which can later be emitted back into space as heat. This emitted energy is known as longwave radiation. Much more heat is emitted from areas near the equator than near the poles. Heat is also carried by ocean currents. Bright colors show areas emitting high amounts of heat; white shows low heat emission. These images are global maps made six months apart showing the amount of longwave radiation detected by the CERES sensor on the Terra satellite.

Unlike multispectral instruments, radar sensors do not detect reflected sunlight. Radar sensors produce their own pulse of energy. This pulse reflects off the surface of the Earth and returns to the satellite. The amount of energy reflected by the surface is measured and recorded, and the data are used to determine characteristics such as the composition and roughness of the surface.

The size of the area on the ground represented by one pixel (called the spatial resolution of a sensor) determines how much detail is visible in the image. On GOES and other weather satellites observing the atmosphere, pixels often represent as much as 1 kilometer (0.6 miles). Satellites like Landsat and Terra, which observe regional-scale patterns, have pixels representing 15–30 meters. Images from the new breed of satellites, which include IKONOS and QuickBird, have resolutions of about one meter, and future satellites will have even higher resolution.

Mapping from Space and on the Ground

Accurate mapping of the Earth's topography is possible from orbit. Some satellites have sensors that tilt, so they can observe the Earth at an angle instead of straight down. This gives them the ability to obtain stereo views. Another way to make topographic maps from orbit is by using radar. Several radar sensors have been orbited, including the Seasat satellite in 1978. The Shuttle Radar Topography Mission (SRTM) in February 2000 returned a huge amount of high-quality data, for the first time allowing the creation of a highly detailed global topographic map.

Another aerospace technology that has provided new ways of looking at the Earth is the Global Positioning System (GPS). GPS is a system of satellites that allows anyone with a hand-held device to determine their location anywhere in the world with great accuracy. GPS was designed and built by the U.S. military to provide positioning data for missiles and solders in the field. Today surveyors use GPS in their work and scientists use GPS and satellite images to navigate around field areas and better understand the Earth's surface.

Viewing the Earth

People who work in the remote sensing field are trained in a variety of subjects. Some satellite missions are purely scientific, while others are commercial and still others are both scientific and commercial. Before the images can be used, analysts process the data. When digital remote sensing began in the 1970s, very large computers were necessary to process satellite images. Since the mid-1990s ordinary desktop computers — not very different from the ones typically found

New Sources of Imagery

As technology has advanced, remote sensing platforms have become smaller and the data have become more widely available. This image of the city of Pusan, South Korea was acquired by EarthKAM, a digital camera on the International Space Station. The camera is controlled by school students over the internet.

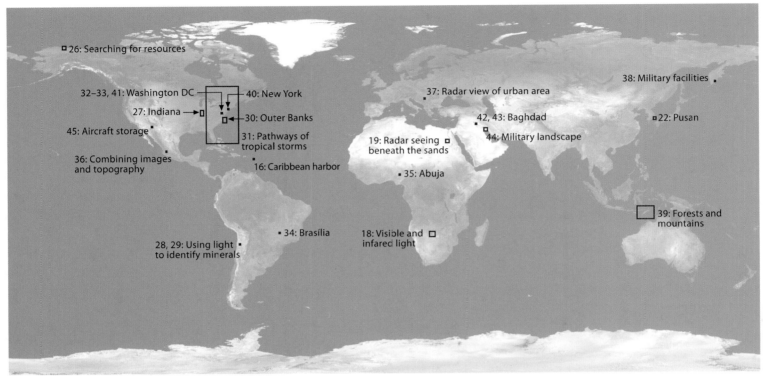

Numbers indicate page locations of the images in this chapter.

in homes — can do the job. However, extensive data storage capabilities are required because of the large file sizes typical of remote sensing data. For instance, anywhere between 400 and 1,500 megabytes are required to store a single image from the Landsat 7 satellite.

When first received on the Earth, satellite images contain errors, do not correspond to any maps and do not show many details. The first step to making satellite images useful is to eliminate errors and calibrate the data. This is done to ensure that the amount of light being measured by the sensor is consistent across the image. The next step is to make the images match maps so that every pixel has a map coordinate. This is done by digitally warping the image to fit onto map coordinates. The contrast of the image can be enhanced to bring out certain details. Finally, the image data are ready for use.

Remote sensing satellites have been launched since the 1970s, but only since about 2000 have image-processing and data-sharing technologies been available to a broad range of people. New satellites and expanded access to technology have helped open doors to a better understanding of the planet we live on.

Orbits

Astronauts circle the Earth in equatorial orbits about 200–300 miles (300–500 km) above the surface. Remote sensing satellites are usually in polar orbits at about 300–500 miles (400–800 km) altitude, traveling north and south over the poles. Most satellites are placed into inclined "sun synchronous" polar orbits, allowing the Earth's surface to be observed at a constant sun angle. Weather satellites are placed into geostationary orbits 22,000 miles (36,000 km) above the Earth's surface. At this distance, a satellite travels around the Earth once every 24 hours, the same speed at which the Earth rotates. Because the satellite stays stationary relative to a point on the surface, it can constantly observe an entire hemisphere.

Remote Sensing Satellite

Remote sensing satellites have solar panels to produce electricity, antennae for transmitting data to receiving stations on the Earth, and sensors for observing the Earth's surface. This diagram shows the Landsat 7 satellite, but most remote sensing satellites have similar components.

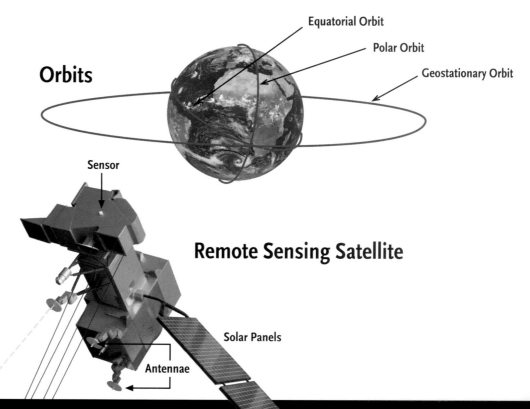

Orbits

Equatorial Orbit

Polar Orbit

Geostationary Orbit

Remote Sensing Satellite

Sensor

Solar Panels

Antennae

Data from Satellite

Receiving Antenna on the Earth

Electromagnetic Spectrum

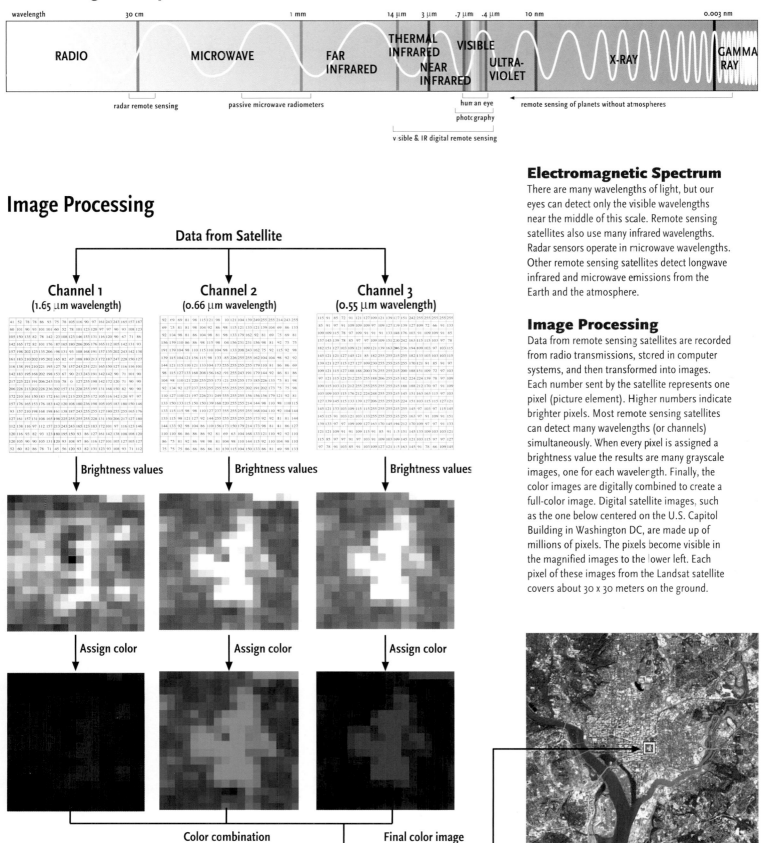

Image Processing

Electromagnetic Spectrum

There are many wavelengths of light, but our eyes can detect only the visible wavelengths near the middle of this scale. Remote sensing satellites also use many infrared wavelengths. Radar sensors operate in microwave wavelengths. Other remote sensing satellites detect longwave infrared and microwave emissions from the Earth and the atmosphere.

Image Processing

Data from remote sensing satellites are recorded from radio transmissions, stored in computer systems, and then transformed into images. Each number sent by the satellite represents one pixel (picture element). Higher numbers indicate brighter pixels. Most remote sensing satellites can detect many wavelengths (or channels) simultaneously. When every pixel is assigned a brightness value the results are many grayscale images, one for each wavelength. Finally, the color images are digitally combined to create a full-color image. Digital satellite images, such as the one below centered on the U.S. Capitol Building in Washington DC, are made up of millions of pixels. The pixels become visible in the magnified images to the lower left. Each pixel of these images from the Landsat satellite covers about 30 x 30 meters on the ground.

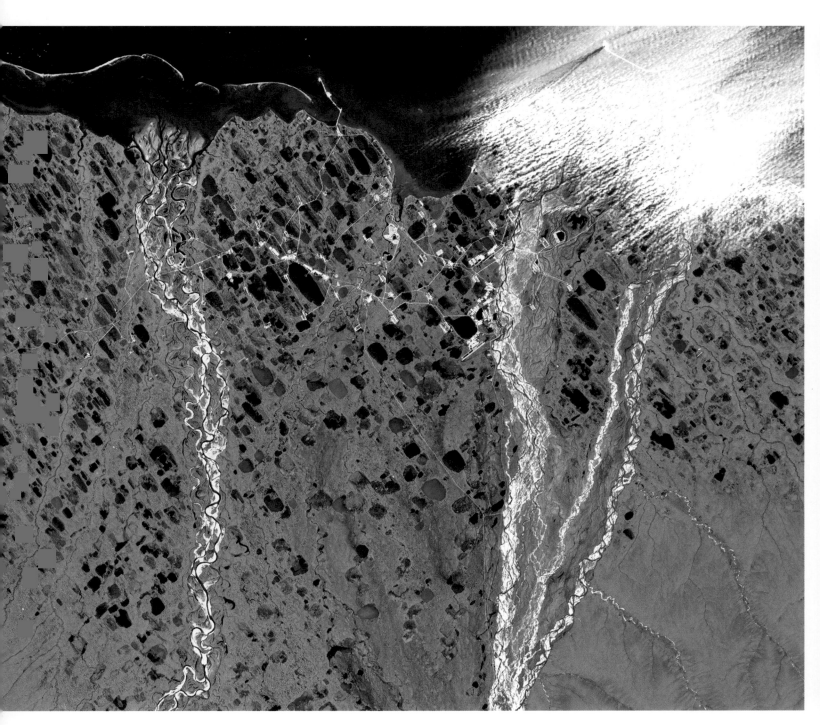

Searching for Resources

PRUDHOE BAY, ALASKA (ABOVE)

Remote sensing data are used to identify geological structures likely to contain resources such as petroleum. This image, collected by the Landsat 7 satellite in August 3, 1999, shows Prudhoe Bay, Alaska. The surrounding area became a focus of national interest in 1968 when petroleum was found here. Several large oil fields were developed, some on land and others at sea. A network of oil drilling equipment is visible near the center of this image. This area accounts for about 20 percent of oil production in the United States. The Trans Alaska Pipeline was completed in 1977 to transport crude oil to ports in southern Alaska. In the 1990s possible petroleum reserves were identified east of Prudhoe Bay within the Arctic National Wildlife Refuge, leading to debate over management of the land.

Changing Views from Space

The image on the right, acquired by the Landsat 7 satellite on July 11, 2001, shows landcover types in central Indiana. The Indianapolis metropolitan area is visible in the upper half of the image and the city of Bloomington lies in the lower left. Digital images such as these are made up of data transmitted to receiving stations on the Earth's surface. The small image below, collected by a Corona camera on May 1, 1965, shows small details in central Indianapolis. Corona was a U.S. reconnaissance program that used film cameras. Unlike today's digital systems, the exposed film was returned to Earth in re-entry capsules.

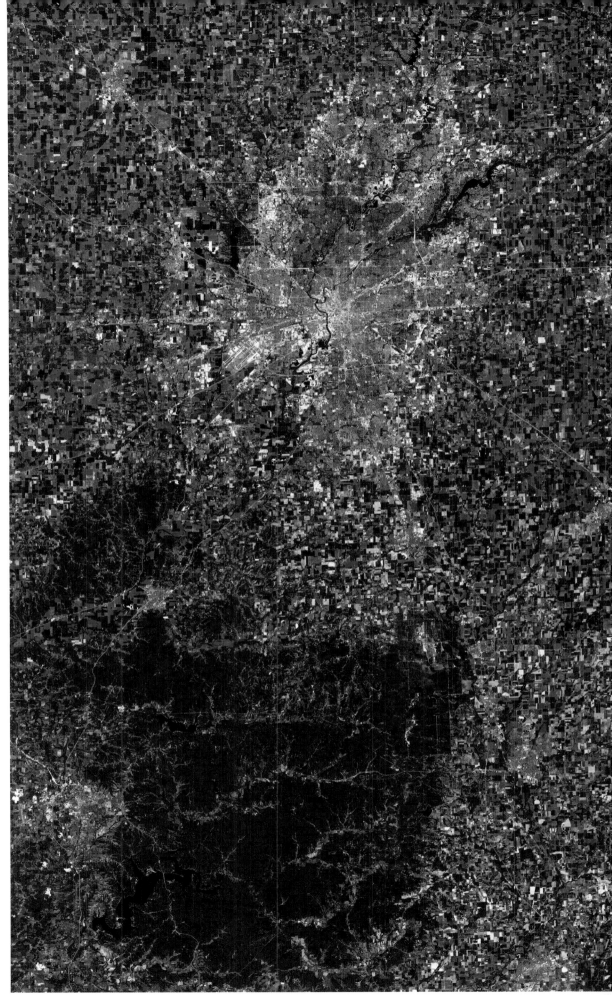

Using Light to Identify Minerals

ESCONDIDA MINE, VISIBLE LIGHT (LEFT)
INFRARED LIGHT (ABOVE)

These images from the ASTER sensor on the Terra satellite show the Escondida mine in northern Chile. The mine is 10,000 feet (3,050 m) above sea level in the Atacama Desert, an area that receives almost no precipitation. Since 1990 the mine has produced mostly copper, but also some silver and gold. Every day, about 127,000 tons of rock are moved into the processing area. The ore is then piped to the Pacific coast through a 100 mile (160 km) pipe. In 1999 the mine yielded 827,000 tons of copper — more than half the total copper mined in Chile.

The image on the left shows the Escondida mine in visible light, as it would appear to our eyes. The image above shows the same area using infrared light. In contrast to the infrared image of the Okavango Delta (page 18), th s image uses longer wavelengths of infrared light. These show differences in geology, making such images useful for detecting minerals and changes created by mining activity.

Pathways of Tropical Storms

OUTER BANKS, NORTH CAROLINA (LEFT)
The image on the left shows the Outer Banks of North Carolina, a string of low-lying islands just off the coast. These barrier islands are more than 100 miles (160 km) long, but often only a few hundred feet wide. Wind and waves are slowly moving the sandy islands closer to the land. Large storms and hurricanes can tear gaps through the islands. The Landsat 7 satellite collected this image on September 23, 1999. It shows the effects of Hurricane Floyd, a storm that flooded rivers and damaged or destroyed about 7,000 homes. Green indicates vegetation on land and phytoplankton in the water.

TROPICAL STORM GUSTAV (RIGHT)
This image of tropical storm Gustav was acquired by the MODIS sensor on the Terra satellite as the storm approached the Outer Banks on September 10, 2002. The coastline has been superimposed on this image to show where the storm is approaching land. Gustav was a powerful storm that could have caused significant damage, but it turned northeast and eventually moved back out into the Atlantic. Weather forecasters use satellite image data to predict where storms will touch land so that people living along the coast may be evacuated.

Planning the City

WASHINGTON DC

This high-resolution image from the QuickBird satellite shows central Washington DC. Imagery acquired at this scale, showing the same detail visible in photographs taken from airplanes, can be used for urban-planning purposes. Washington, a planned city, was created in the 1790s as the new capital of the United States. Engineer and architect Peter L'Enfant planned the urban layout of the gridded streets and broad diagonal avenues still visible today. Near the

center of the image is the 555 foot (170 m) Washington Monument, casting a shadow to the northwest. The Lincoln Memorial lies to the west of the monument. This area was originally submerged before the Potomac River was dredged to create new parkland. The U.S.

Capitol lies to the east. Stretching across the center of the image is the National Mall, with the Smithsonian museums on either side. To the northeast is Union Station, the largest train station in the world when it opened in 1907.

Planning the City

BRASÍLIA, BRAZIL (ABOVE)

ABUJA, NIGERIA (RIGHT)

These high-resolution images show two cities deliberately planned and created as national capitals. The image above shows Brasília, the capital of Brazil. The site was chosen for a new capital in 1956, and in 1960 the government started to move from Rio de Janeiro. The area's population grew from 60,000 in 1960 to almost 2 million in 2000. The layout of the city resembles a bird, as can be seen in this image from the QuickBird satellite. Large government buildings are near the Monumental Axis, which stretches from east to west across the center of the city. Stretching north and south is the Highway Axis, the "wings" of the bird, where residential areas are located. Part of artificial Lake Paranóa is visible to the east.

On the right is an image from the IKONOS satellite showing Abuja, the capital of Nigeria. Extensive construction began here in the 1980s, and in 1991 Abuja became the seat of government. Today, more than 500,000 people live in Abuja. Like Brasília, the site was chosen because it was near the center of the country. Many government buildings, including the National Assembly, are visible near the center. Lagos, the previous capital, is Nigeria's largest city, with more than 13 million people.

Combining Images and Topography

GUADALAJARA (LEFT)

The image on the left is a perspective view created by computer software that combines Landsat 7 image data with elevation data. More than 400 years old, Guadalajara is the second-largest city in Mexico and has a metropolitan population of over 2 million people. Guadalajara lies in a valley about 5,300 feet (1,600 m) above sea level. On the left is Lake Chapala. The high mountain in the background is Nevado de Colima. Topography has been exaggerated in this view.

Radar Views of Urban Areas

BELGRADE (ABOVE)

In October 1994, the Space Shuttle carried a radar sensor on the SIR-C mission. This SIR-C image shows the city of Belgrade, the capital of the republic of Serbia, near the confluence of the Danube (running top to bottom) and Sava rivers. The central part of the city where the rivers join is visible here. Large angular structures such as buildings strongly reflect radar energy, making urban areas show up very brightly in radar images. Urban surfaces appear in yellow tones, and individual large buildings are clearly visible. Agricultural fields are dark blue, and forests in the lower left are dark green.

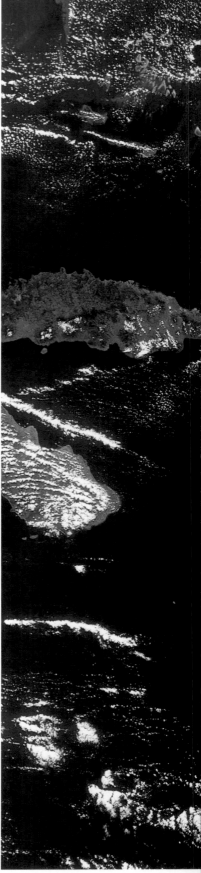

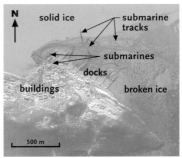

solid ice
submarine tracks
submarines
docks
buildings
broken ice
500 m

Military Facilities

KAMCHATKA PENINSULA (ABOVE)

The home port of the Russian Pacific
submarine fleet on the Kamchatka
Peninsula is located on Avacha Bay
south of the city of Petropavlovsk.
Fifteen submarines are docked at the
port, which is visible in the center of
the image. The upper part of the image
shows a sheet of solid ice. A path
through the ice ends where a departing
submarine submerged. This image,
acquired by the EROS A satellite on
December 25, 2001, is similar to those
returned by reconnaissance satellites.

Forests and Mountains

ISLAND OF TIMOR (RIGHT)

The island of Timor is visible in this
image from the MODIS sensor on the
Terra satellite. The central part of the
island is mountainous, with the tallest
peak almost 10,000 feet (3,000 m) above
sea level. The high rainfall allows dense
vegetation to thrive in the central parts
of the island. Western Timor is part of
Indonesia, but the rest of the island is
the independent nation of East Timor.
Active volcanoes can be seen on the
island of Flores, just north of Timor.
A phytoplankton bloom is visible to the
southeast. This type of image is used to
monitor environmental changes on the
Earth's surface.

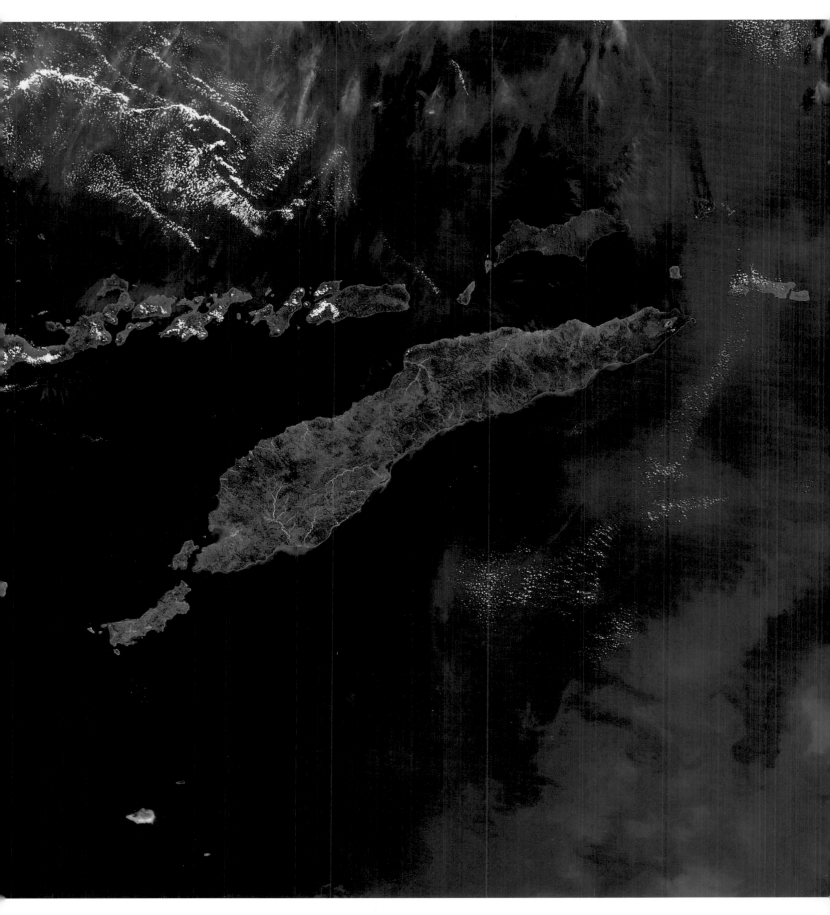

World Trade
Center site

500 m

Planning and Rebuilding

NEW YORK (LEFT)

The image on the left shows parts of New York City and the skyscrapers that dominate lower Manhattan, the world's most important financial center. The towers of the World Trade Center, over 1,300 feet (400 m) high, were destroyed by the terrorist attack of September 11, 2001. Visible within the site where the World Trade Center stood is a subway station, which resumed operations in November 2003. The rest of the site will contain new skyscrapers and a memorial to those who died. High-resolution satellite images such as this one, acquired by the QuickBird satellite in August 2002, can be used for planning future construction.

THE PENTAGON (RIGHT)

The image on the right shows the Pentagon, the headquarters of the U.S. military, just outside Washington DC. The Pentagon was built between 1941 and 1943. When completed, it was the world's largest building, with 15 miles (24 km) of hallways and room for 25,000 people. Terrorists crashed an airplane into the upper-right side of the building on September 11, 2001, coincidentally the 60th anniversary of the Pentagon's groundbreaking. Despite the damage the building never closed, and the damaged section was completely rebuilt within one year. This image from the IKONOS satellite shows ongoing reconstruction two months after the attack.

Modern Warfare

BAGHDAD AIR STRIKES (ABOVE)

The effects of modern warfare are clearly visible in the image above, acquired by the IKONOS satellite. When United States forces attacked Iraq in April 2003, air power was used extensively to destroy government buildings. A presidential palace in Baghdad is shown here after being hit by air strikes. The roof of the building has been destroyed.

FIRES AROUND BAGHDAD (RIGHT)

The IKONOS image on the right shows an overview of Baghdad. During the attacks the Iraqi military set large fires in an unsuccessful attempt to confuse aircraft and missiles. Large plumes of smoke are visible from the fires.

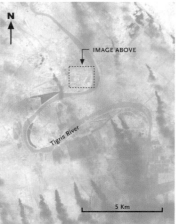

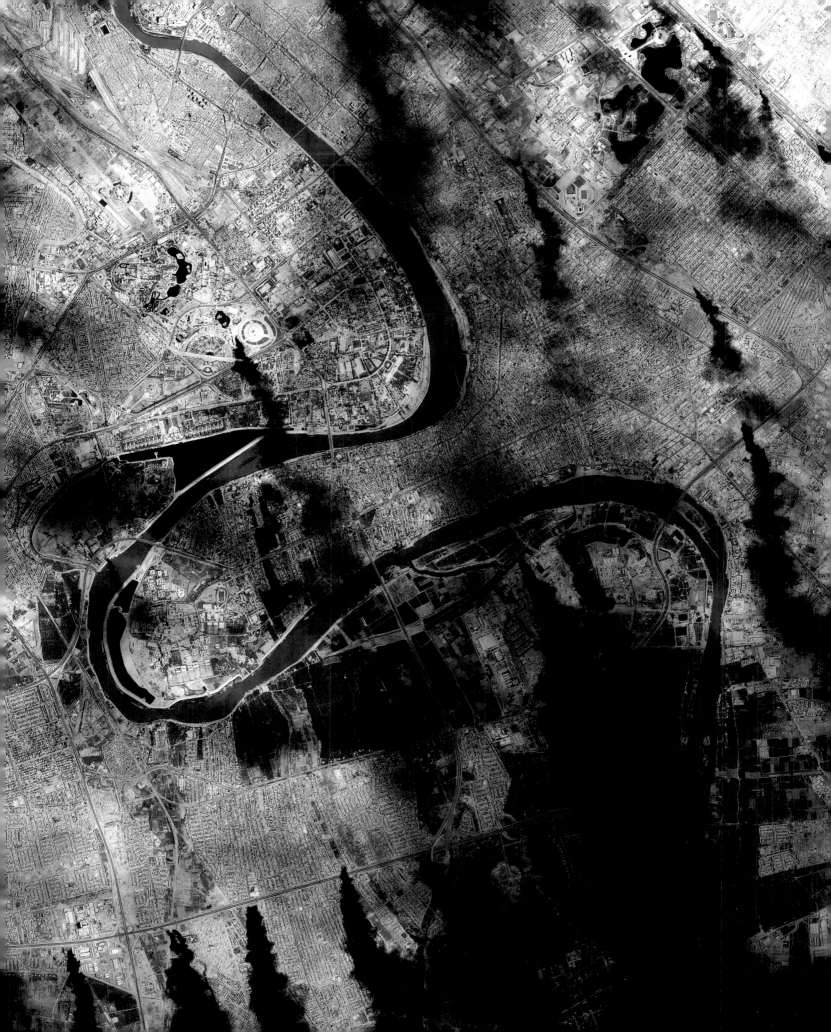

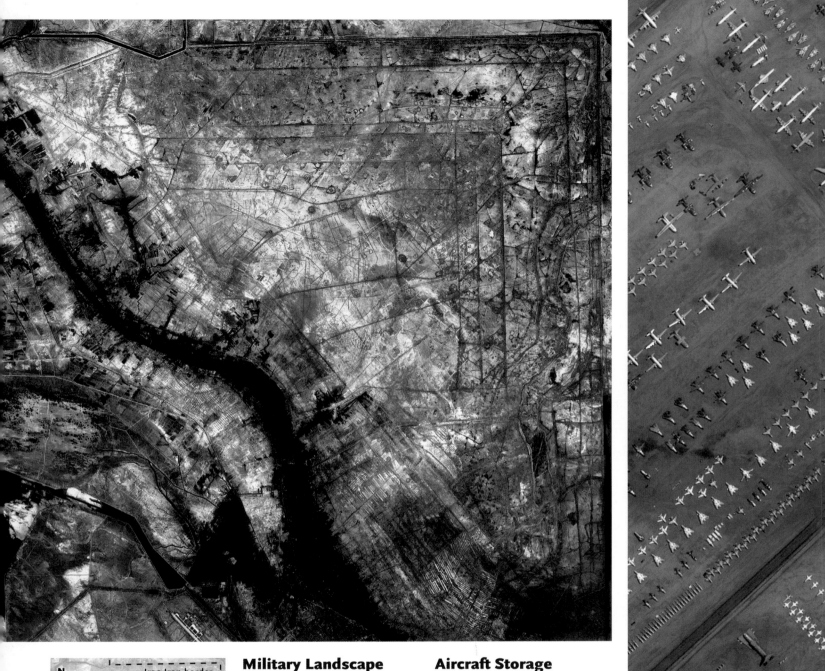

Military Landscape

IRAQ-IRAN BORDER (ABOVE)

This image, acquired by the Landsat 7 satellite in January 2001, shows the border area between Iraq and Iran northeast of the Iraqi city of Basrah. The straight lines show fortifications built by Iraq during the war with Iran in the 1980s. Much of this area was a wetland marsh, but it was drained during the 1990s.

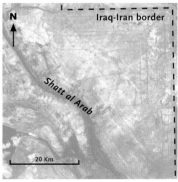

Aircraft Storage

DAVIS-MONTHAN AIR FORCE BASE (RIGHT)

Davis-Monthan Air Force Base near Tucson, Arizona, holds obsolete and surplus aircraft until their ultimate fate is decided. The dry desert air of southern Arizona keeps the aircraft free of rust. B-52 bombers lying with their detached wings are visible in the lower-right part of this image from the QuickBird satellite.

Living Planet

bewildering diversity of life exists all over our world. Life in the ocean ranges from microscopic organisms that float near the surface to creatures that live under immense pressure in the perpetual darkness of the ocean's depths. On land, life can be found in every climate and environment — from deserts to tropical forests, from mountaintops to deep canyons. Living things interact with each other by exchanging nutrients and gases in a complex web of ecosystems often called the Earth's biosphere.

Satellites in low orbit observe plant life on the land, the density of organisms in ocean water, and the interaction between living things and the atmosphere. Other satellites track short- and long-term changes in the atmosphere. The images from these satellites provide insight into changes taking place on our living planet.

Forest Ecosystems

Wherever enough moisture, soil and sunlight are available, plants will find a way to grow. Forest ecosystems cover about a third of the Earth's land surface. The forests are filled with a wide diversity of trees. Coniferous trees, with layers of long needles, stay green all year. Deciduous trees lose their larger, flat leaves during the winter in cooler climates.

Boreal forests, dominated by low-density stands of coniferous spruce trees, occupy a belt around the northern regions of North America, Europe and Asia.

Temperate forests cover areas where temperatures are warmer, the growing season is longer and adequate moisture is available for tree growth. These forests contain a mixture of broadleaf deciduous and coniferous trees.

Tropical forests dominate the Earth's equatorial zones, where temperatures

River of Grass
THE EVERGLADES
The Florida Everglades, opposite, was created by water flowing out from Lake Okeechobee toward the southern tip of Florida. Most of the water is between 1 and 3 feet (30 and 90 cm) deep, and it flows about 100 feet (30 m) per day. The Everglades is sometimes called the "river of grass" because of the huge amount of grasses that grow in the wet soil there. The Everglades forms an important wildlife habitat. It is the only place in the world where both alligators and crocodiles live together. Everglades National Park protects the area against development from nearby Miami, to the east. Cutting across the image is the Tamiami Trail, a busy highway running west from Miami. This Landsat 7 image was collected in February 2000 during the dry season. The complex colors indicate different types of vegetation.

are warm all year and rain falls frequently. Characterized by dense stands of tall broadleaf trees, tropical forests are particularly important reservoirs of life and contain more than 50 percent of the Earth's plant and animal species. This band of dense forests exists around the equator in South America, Africa and Asia.

Wetlands and Drylands

Wetlands exist where the land surface is inundated by water. There are several types of wetlands, called fens, bogs, marshes and swamps, but they are all rich in plant life and have abundant water. Grassy vegetation can also thrive in places where ocean water inundates the land, creating saltwater wetlands.

About 40 percent of the Earth's surface is covered by drylands. These are areas that receive less rainfall than forested areas do, but sufficient moisture to support many species of grasses and shrubby vegetation. Drylands go by many names: plains, grasslands, savannas, steppes and pampas. The plains of North America are covered by vast grasslands to the foothills of the Rocky Mountains. The savannas of Africa are home to the world's largest populations of grazing animals.

Observing Plant Life from Space

Some wavelengths of light are absorbed by plant leaves while others are reflected back into the sky. The wavelength of light that we see as the color red is absorbed more strongly by plants than is the wavelength that we see as green. That is why leaves appear green: we are seeing the reflected green light. Plant leaves also reflect in the near-infrared part of the spectrum. To be able to identify vegetation cover, satellites such as Landsat were designed with these characteristics in mind.

Scientists use remote sensing data to observe plants in many ways. One method involves measuring the difference between reflected near-infrared light and red light. Because plants reflect near-infrared light and absorb red light, this measurement, called a vegetation index, is often used to help scientists obtain estimates of vegetation growth.

Scientists can estimate the density and number of trees in a forest by creating mathematical models of how trees interact with sunlight. Some of these models assume that forest canopies consist of layers of vegetation which permit sunlight to pass through them. Other models assign abstract shapes to different trees. The most complicated models attempt to describe how forests reflect light according to the density and orientation of millions of individual leaves. The surface area of leaves relative to the area of Earth's surface is known as the leaf-area index. Many

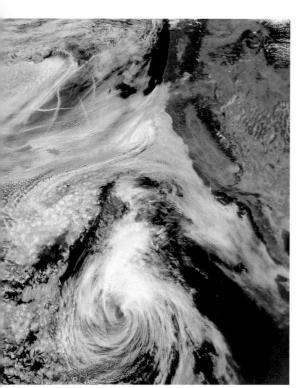

Clouds over the Pacific

This image shows swirling clouds and smoke over the Pacific Ocean off the coast of California. Tropical storm Elida lies in the southern part of the image. To the north of the storm are plumes of smoke that have a dark yellowish color. The smoke, coming from forest fires in California and Oregon, has drifted far out into the Pacific. North of the smoke, straight lines are visible. These are ship tracks caused by the exhaust from ships traveling the seas. This image was acquired by the SeaWiFS sensor on the OrbView-2 satellite in July 2002.

techniques using images from space have been developed to estimate this index, which plays an important part in helping scientists understand how trees are structured.

The Earth's Atmosphere

The atmosphere makes life possible on Earth. Several other planets have atmospheres, but one similar to the Earth's has not yet been found. The blanket of air surrounding the Earth protects the surface from harmful radiation and allows plants and animals to breathe. Molecules of several gases mix together in the turbulent atmosphere. The abundance of these molecules is difficult to imagine, but the average breath contains about the same number of molecules as there are stars in the universe.

Our atmosphere is about 78 percent nitrogen, a transparent, inert gas. Most of the rest (about 21 percent) is oxygen, which was created by plant life. Plants take in carbon dioxide and release oxygen. Billions of years ago, before plant life existed, there was a negligible amount of oxygen in the atmosphere. About 0.03 percent of today's atmosphere is made up of carbon dioxide. The abundance of this gas is low because it is continuously absorbed by plant life. Most of the remaining 1 percent of the atmosphere is the gas argon. Argon is created from the radioactive decay of the element potassium in the Earth's crust. The abundance of water vapor in the atmosphere varies between 0 and 4 percent, depending on weather conditions.

The atmosphere is made up of four layers. The troposphere extends from the Earth's surface to an altitude of about 10 miles (17 km). All the weather we experience happens within this layer. Next is the stratosphere, reaching about 28 miles (48 km) above the surface. In the upper parts of this layer the air is very thin and has a higher temperature, which is caused by the absorption of ultraviolet solar radiation. Above the stratosphere are the mesosphere and the thermosphere, and then the atmosphere finally thins out to almost nothing at an altitude of 80 miles (140 km).

One of the trace gases in the atmosphere is ozone, a form of oxygen with three atoms instead of the usual two. At low altitudes ozone is a pollutant that irritates lungs and is harmful to plants. However, about 15 miles (25 km) into the stratosphere ozone is a highly useful gas. Ozone protects the Earth from the Sun's harmful ultraviolet radiation. The ozone layer was damaged by chemical pollution in the 1900s, but has partially recovered after the chemicals (chlorofluorocarbons, or CFCs) were banned in the 1990s.

Gulf Stream in the Atlantic

The Gulf Stream, a current of warm sea water, is visible here as it moves northward in April 2003. This SeaWiFS image has been color-coded to reveal concentrations of chlorophyll in the ocean. Reds and yellows indicate high levels of chlorophyll, found in microscopic marine plants called phytoplankton. The bright red and yellow zone shows an area of cool water at the northern edge of the Gulf Stream. Phytoplankton thrives better in cooler water, so chlorophyll concentrations fall abruptly at the edge of the warm water. As it moves, the Gulf Stream leaves behind swirling pools of warmer water, one of which is visible as a yellow circle far into the Atlantic. These swirls can survive for over a year.

Watching the Clouds

Perhaps the most widespread application of remote sensing is weather tracking. Before the arrival of satellites, predicting the weather was a guessing game. Nobody could accurately predict when storm systems were approaching from the ocean.

Some of the earliest satellites were intended to monitor the weather. The TIROS satellites, first launched in 1960, were placed into low orbits and equipped with television cameras to record the movement of clouds. TIROS was followed by other satellites with more capable cameras. Today, a fleet of satellites in similar orbits observes weather systems at a regional scale.

Another class of satellites is placed in distant orbits to monitor the entire planet and observe global-scale weather patterns. These satellites travel in geostationary orbit (see diagram on pages 24–25) and are able to observe an entire hemisphere. There are several satellites in such orbits, some operated by U.S. government agencies and others by European, Russian and Japanese agencies. They provide the most well-known satellite images, familiar to anyone watching weather reports on the television news.

Weather satellites also observe the atmosphere with multispectral sensors called sounders. These instruments observe infrared wavelengths that are emitted and scattered by gases in the atmosphere. By knowing the characteristics of each gas, scientists can determine how much air the sounder is "seeing" through. This allows analysts to measure temperature and the presence of carbon dioxide, ozone and water vapor at different levels in the atmosphere.

Water vapor in the atmosphere is essential to life on the Earth. Water forms clouds, which release rain and snow to deliver water to the land. Global maps of water vapor density can be created and updated frequently using satellite data. Patterns of precipitation can also be identified from space. Some satellites are able to detect falling rain by using radar. Other satellites are able to map areas where snow has fallen.

Storms

As weather systems move through the atmosphere, they often form huge storm patterns. The largest storms on the Earth are hurricanes and typhoons. Each year the Earth experiences between 50 and 100 of these storms which, by transporting warm moist air away from the tropics, have important effects on global climate. These storms, which can be 1,000 miles (1,600 km) wide, create widespread devastation with winds greater than 73 miles (117 km) per hour. Hurricanes form in tropical areas, gaining power as they move over warm water.

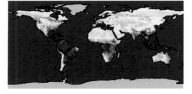

Vegetation on the Land

Green leaves and gases contain chlorophyll, which interacts with sunlight in a specific way. Chlorophyll absorbs red light more than it absorbs green light. Plant leaves also reflect near-infrared light very strongly. These characteristics were used by the AVHRR sensor on NOAA satellites to detect the chlorophyll contained in vegetation, creating the maps shown here. Dark colors indicate a large amount of vegetation, while the bright colors indicate a lack of vegetation. These two maps were made six months apart to indicate the difference between summer and winter.

Even a small storm contains a huge amount of energy. The average thunderstorm releases more energy than the largest power plants, and a single tornado generates enough power to supply more than 10 million homes. Tornadoes form when fast-moving air in a thunderstorm starts to rotate, creating highly destructive winds. Tornadoes most frequently occur in the central United States, which experiences more than 700 every year. A tornado's path of destruction can be seen and mapped from orbit.

Atmospheric Circulation

The circular shape of hurricanes and other weather systems on a spinning body like the Earth is determined by the Coriolis effect. This effect causes moving air to be deflected to the right in the Northern Hemisphere and to the left in the Southern Hemisphere. This makes large circulation patterns move clockwise in the north and counterclockwise in the south. It is often said that water spins around the drain in opposite directions in different hemispheres. In fact, this is not true. The Coriolis effect is a factor only over large areas — much larger than the width of a sink.

Global-scale air circulation patterns are also visible from space. As the surface of the ocean and land are warmed by the Sun, this warmth is passed to the atmosphere. The warm air rises and travels north and south until it cools and falls back toward the surface. These circulation patterns, called Hadley cells, form huge belts of higher and lower precipitation reaching completely around the planet. The largest area of high rainfall is the intertropical convergence zone near the equator, which creates the Earth's tropical forests.

Oceans

Oceans cover about 70 percent of the Earth's surface, contain about 320 million cubic miles (1.3 billion cubic km) of water, and control the global climate. About 98 percent of the Earth's water is stored in the oceans. A cubic mile of seawater — a concoction of water, salts and minerals — contains 130 million tons of sodium chloride (table salt) and about 12 million tons of other compounds and elements.

Microscopic organisms, called plankton, dominate life in the oceans and form the basis of the ocean food chain. One type of organism, known as phytoplankton, uses chlorophyll to absorb sunlight to fuel growth. In the process, phytoplankton absorb carbon dioxide and create about half the oxygen in the atmosphere. Phytoplankton usually live within 35 feet (10 m) of the water surface. They are visible from space with the aid of sensors on satellites such as Terra and Aqua.

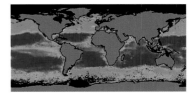

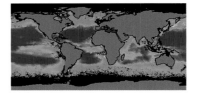

Life in the Oceans

These global maps show the presence of life in the Earth's oceans in different seasons. Phytoplankton — marine plant life floating in the oceans — use chlorophyll to absorb the sunlight needed for growth. The basis for the entire ocean food chain, phytoplankton also absorb carbon dioxide and creates about half the oxygen in our atmosphere.

Specific wavelengths of sunlight are used by the MODIS sensor on the Terra satellite to detect the presence of phytoplankton. This information is important for understanding the impact sea life has on global climate. The bright colors on these global maps show high concentrations of chlorophyll. These two maps were made six months apart.

The oceans play an important role in global climate. Although air temperatures can change rapidly, it takes far greater energy to heat the huge mass of water in the oceans. It takes less than 10 feet (3 m) of sea water to hold the same amount of heat energy found in the entire atmosphere above it. That makes the oceans able to absorb thermal energy and slow down the temperature changes in the atmosphere. The oceans also contain vast amounts of carbon dioxide, about 50 times as much as exists in the atmosphere, which may have a significant effect on global climate. Ocean currents move huge amounts of warm water from the equatorial regions to polar extremes, warming vast areas.

The Gulf Stream, first mapped by Benjamin Franklin in 1769, is a current of warm water that flows to the northeast along the east coast of North America. Where a branch of the Gulf Stream flows by Florida, it moves at its fastest speed: about 4 miles (6.5 km) per hour. The Gulf Stream is often 20°F (12°C) warmer than ocean water just to the north. Eventually the waters of the Gulf Stream warm northwestern Europe. Without the Gulf Stream, air temperatures off the coast of Norway would drop about 40°F (22°C). The Gulf Stream keeps temperatures above freezing in Iceland, makes it possible to grow fruits and vegetables in Britain, and keeps the Russian port of Murmansk on the Arctic Ocean free of ice all year.

Sometimes the ocean's cycles can have drastic effects on global climate. In the southern Pacific Ocean warm water around Indonesia can move across the world to the coast off South America and back again. This process is called the El Niño Southern Oscillation, or ENSO. When warm water arrives it has a devastating effect on the usually rich fishing areas off South America because fewer nutrients can live in warm weather. Because it often arrives in South America around Christmas, fishermen in Peru called it El Niño, or "the (Christ) child." Besides depressing the fishing industry, El Niño affects global climate. Rainfall across North America is changed, and the weather in southeast Asia is affected. When the warm water "sloshes" back across the Pacific, the opposite situation — called La Niña — occurs.

Satellites have been watching the oceans for decades. One of their most important tasks is to take measurements of the temperature of the sea surface. Several satellites carry sensors that can detect infrared wavelengths, which are used to make daily maps of how ocean waters heat and cool the globe.

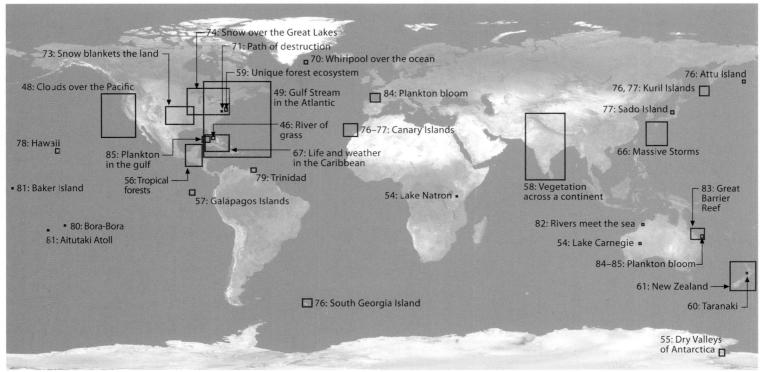

74: Snow over the Great Lakes
71: Path of destruction
73: Snow blankets the land
70: Whirlpool over the ocean
59: Unique forest ecosystem
76: Attu Island
48: Clouds over the Pacific
49: Gulf Stream in the Atlantic
84: Plankton bloom
76, 77: Kuril Islands
77: Sado Island
46: River of grass
76–77: Canary Islands
78: Hawaii
85: Plankton in the gulf
67: Life and weather in the Caribbean
66: Massive Storms
81: Baker Island
56: Tropical forests
79: Trinidad
57: Galápagos Islands
54: Lake Natron
58: Vegetation across a continent
83: Great Barrier Reef
80: Bora-Bora
81: Aitutaki Atoll
82: Rivers meet the sea
54: Lake Carnegie
84–85: Plankton bloom
61: New Zealand
60: Taranaki
76: South Georgia Island
55: Dry Valleys of Antarctica

Numbers indicate page locations of the images in this chapter.

Life Where the Ocean Meets the Land

Among the most diverse places for sea life are reefs, which are in shallow waters near the coastlines of continents and islands. Coral consumes algae and plankton that float by in the water. Individual coral organisms are known as polyps, millions of which grow adjacent to one another and form large colonies.

Coral builds a hard, calcium-rich cover around itself. Mollusks and other animals also leave their shells behind. As time passes, tons of this hard material form layers of limestone to build reefs. Reefs grow as much as 8 inches (20 cm) every year, and some of the largest reefs are less than 10,000 years old. Reefs cover more than 220,000 square miles (570,000 sq km) around the Earth's surface. They are built in places where oceans are shallow, waters are calm, and temperatures range between 65 and 85°F (18 and 29°C). Besides coral, fish and a wide variety of other sea life use coral reefs as a habitat.

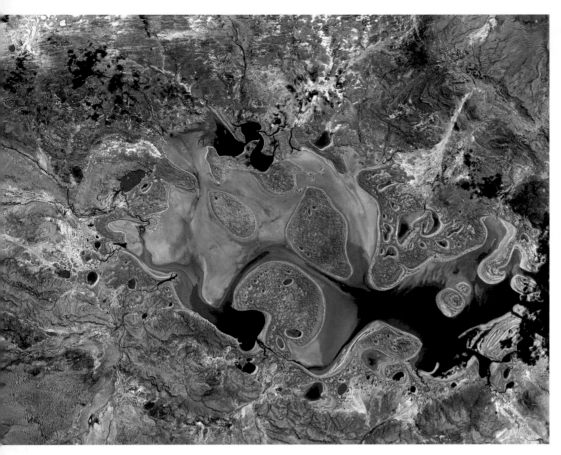

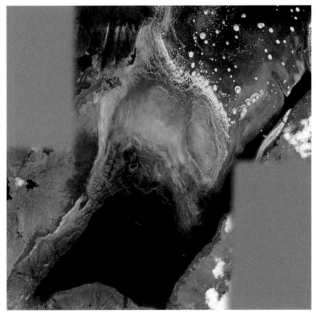

Life in Harsh Environments

LAKE CARNEGIE, AUSTRALIA (ABOVE)

Life exists everywhere on the Earth, even in places that seem totally inhospitable. The image above, acquired by the Landsat 7 satellite on May 19, 1999, shows Lake Carnegie in western Australia. Only prolonged rainfall can make the water level rise in this arid part of the world, and lakes often dry up completely. In this environment microorganisms survive by remaining dormant during dry years. Blue colors indicate areas covered by shallow water.

LAKE NATRON, TANZANIA (LEFT)

This image shows Lake Natron in Tanzania along the East African Rift Valley. Hot springs created by volcanic activity lie beneath the lake, raising its temperature and making the water very alkaline. Mud near the water can reach temperatures as high as 122°F (50°C) and the pH of the water can reach 10.5, almost the same as pure ammonia.

Although Lake Natron is a harsh environment, life thrives there. Microorganisms survive by consuming the alkali salt crust left behind by evaporating water. These organisms are classified as blue-green algae, but they are actually pinkish because of the pigments they carry. Lake Natron is the world's only breeding ground for millions of lesser flamingos, which feed on the algae. The image is made up of two photographs taken by astronauts on the International Space Station in November 2002.

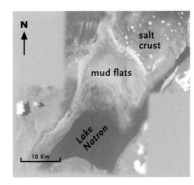

N

salt crust

mud flats

Lake Natron

10 Km

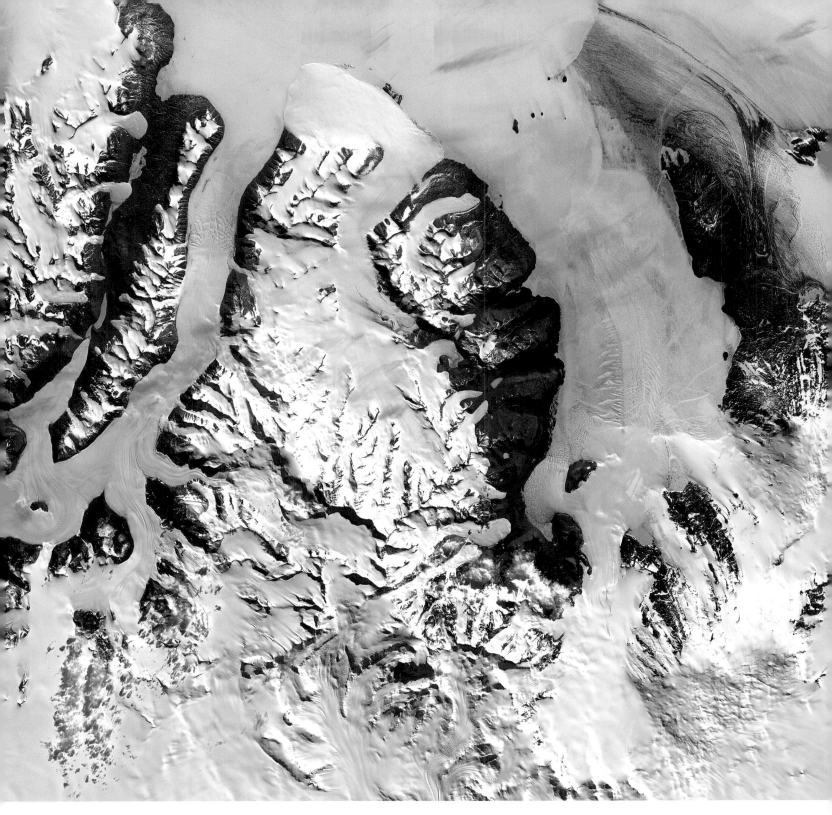

DRY VALLEYS OF ANTARCTICA (ABOVE)

Life can also survive in very cold places. In the McMurdo Dry Valleys of Antarctica, life survives on barren rocks and in lakes that lie beneath layers of ice. The valleys are one of the few parts of Antarctica where rock is exposed rather than being covered by ice and snow. During the summer, temperatures in Antarctica are briefly high enough to melt glacial ice. The melt water drains into lakes at the bottom of the valleys. Although they are covered by an ice cap that is frozen solid, the lakes stay liquid all year and microorganisms thrive in the water. This image was made by the Landsat 7 satellite in December 1999.

Scientists have been studying the Dry Valleys to better understand how life may exist on other planets. With its cold, dry air and melted ice, the environment of the Dry Valleys may be similar to that which existed in the past on Mars. Comparisons have also been made between the Dry Valleys and possible forms of life on Europa, a moon of Jupiter.

Tropical Forests

YUCATÁN PENINSULA (LEFT)

Much of the Yucatán Peninsula is covered by dense tropical forests. The northern part of the Yucatán belongs to Mexico while the southern part includes much of Belize and Guatemala. Also shown here are El Salvador and Honduras. The border between Mexico and Guatemala is especially visible because of the two countries, Mexico's forests have been more extensively thinned.

Fires often break out in the Earth's tropical forests. At any given time, there are many large fires burning in the Yucatán. In this image clouds of smoke are visible. The fires themselves are indicated by small red marks on the satellite image. The environment is dry in the northern part of the Yucatán, but rainfall is more widespread to the south. This allows dense forests to grow, with a wide range of trees and wildlife. Images such as this one, acquired by the MODIS sensor on the Terra satellite on April 2, 2002, show the extent and density of forests.

Islands of Diversity

GALÁPAGOS ISLANDS (ABOVE)

The Galápagos Islands are a unique reservoir of biological diversity made famous by Charles Darwin, who traveled there in 1835. Located about 620 miles (1,000 km) west of South America in the Pacific Ocean, their isolation has limited the number of predators and allowed many species to develop. Each island also has its unique species, including penguins and flamingos. The Galápagos tortoise is a famous resident. On one island there is only one surviving member — named "Lonesome George" — of a tortoise subspecies. Scientists have attempted to get him to mate with a female tortoise from another island, but without success.

Since 1935 the Galápagos have been a protected nature reserve, and today they are one of Ecuador's national parks. This image was acquired by the MISR sensor on the Terra satellite in October 2001.

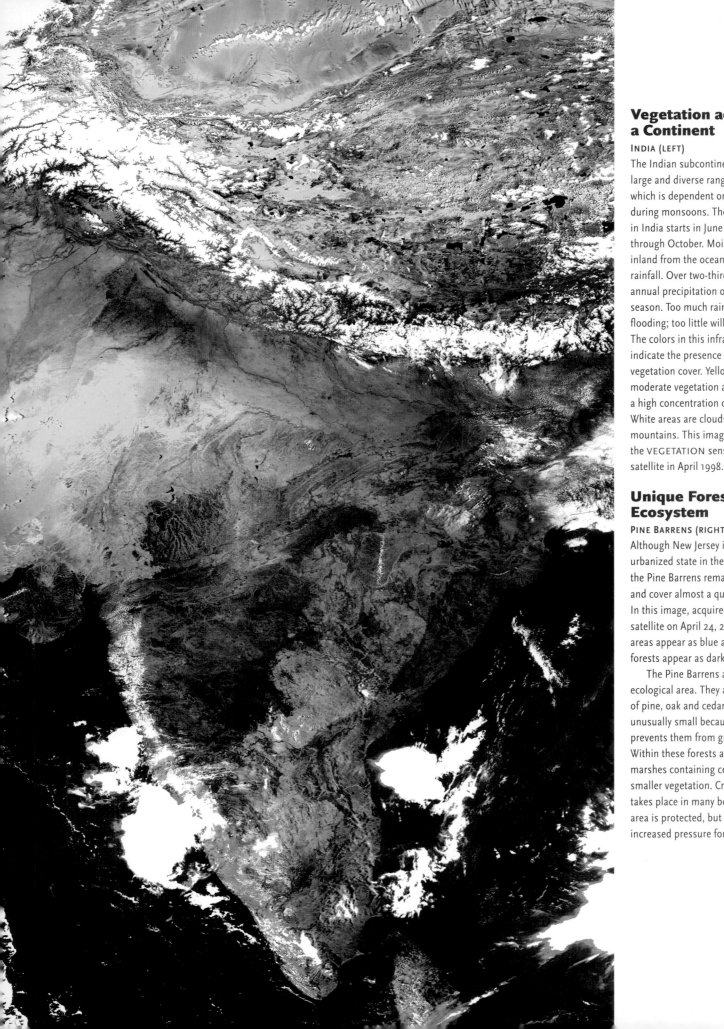

Vegetation across a Continent

INDIA (LEFT)

The Indian subcontinent is home to a large and diverse range of vegetation, which is dependent on the rain that falls during monsoons. The monsoon season in India starts in June and continues through October. Moist winds blow inland from the ocean, causing intense rainfall. Over two-thirds of India's annual precipitation occurs during this season. Too much rainfall can cause flooding; too little will dry up the land. The colors in this infrared image indicate the presence and density of the vegetation cover. Yellow indicates moderate vegetation and red indicates a high concentration of plant life. White areas are clouds or snow-capped mountains. This image was acquired by the VEGETATION sensor on the SPOT 5 satellite in April 1998.

Unique Forest Ecosystem

PINE BARRENS (RIGHT)

Although New Jersey is the most urbanized state in the United States, the Pine Barrens remain densely forested and cover almost a quarter of the state. In this image, acquired by the Landsat 7 satellite on April 24, 2002, urbanized areas appear as blue and purple, while forests appear as dark green.

The Pine Barrens are a unique ecological area. They are made up of pine, oak and cedar trees that are unusually small because the sandy soil prevents them from growing very large. Within these forests are low-lying marshes containing cedar trees and smaller vegetation. Cranberry cultivation takes place in many bogs. Much of the area is protected, but the land is under increased pressure for development.

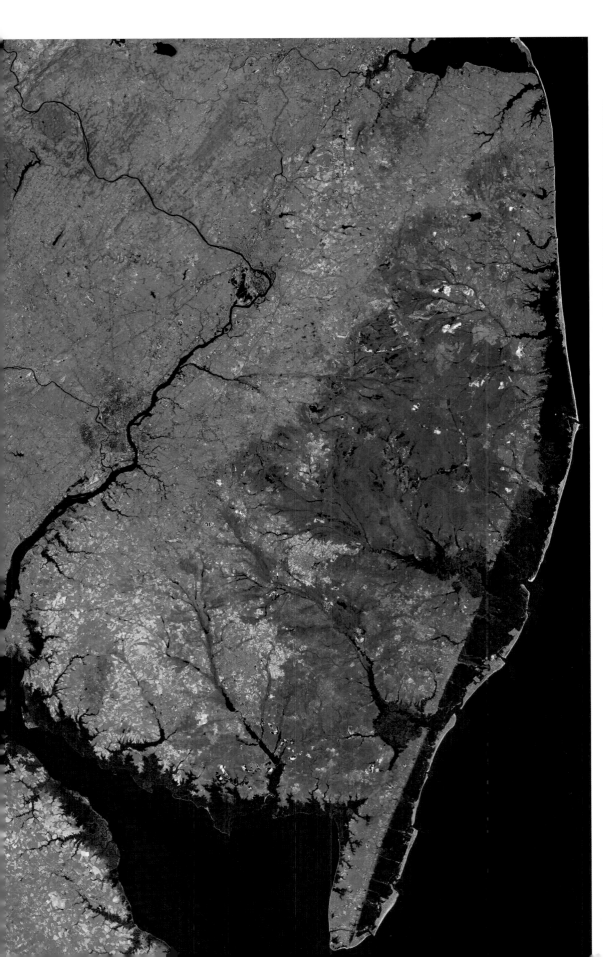

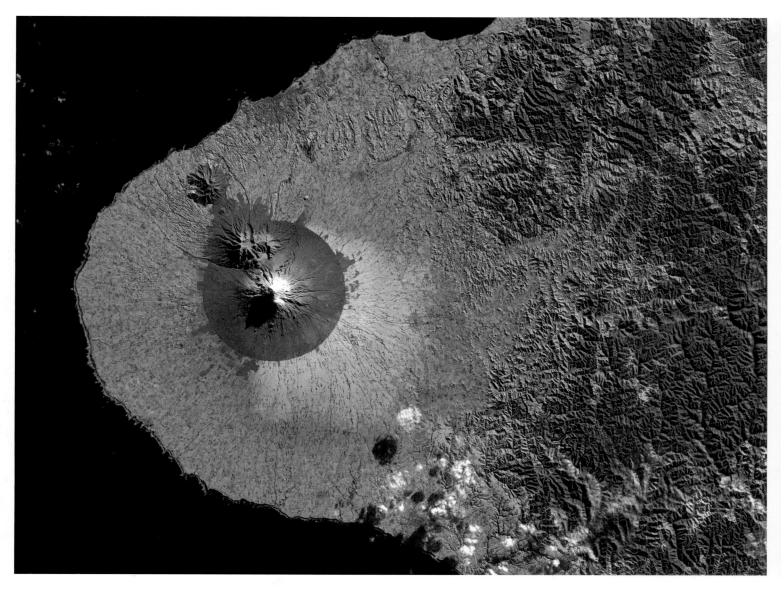

Complex Ecosystems

TARANAKI, NEW ZEALAND (ABOVE)

On the North Island of New Zealand, Taranaki (also called Mount Egmont) is a large volcanic peak on the Taranaki Peninsula, surrounded on three sides by the waters of the Tasman Sea.

Two other older volcanic peaks, Kaitake and Pouaki, lie to the north. All three volcanoes are inside Egmont National Park, which was established in 1900. The circle surrounding the mountain has a radius of 5.7 miles (9.6 km). It is the border of a forest reserve, created in 1881, which has been protected from human use. The land outside the park is used primarily for agriculture. The circle is visible because this Landsat 5 image from June 20, 1989, was acquired immediately after a light snowfall blanketed the ground. The snow is plainly visible on the fields outside the park, but trees obscure the snow that has fallen inside.

NORTH AND SOUTH ISLAND (RIGHT)

New Zealand consists of the North Island and the South Island. Both islands have large mountain ranges crossing from north to south, with 16 peaks taller than 10,000 feet (3,000 m). More than 350 glaciers are in the mountains of the South Island, and the North Island has many volcanic features. Perhaps no other place in the world has such a variety of environments in so small an area. This image was acquired by the MODIS sensor on the Terra satellite on December 31, 2002.

Most of New Zealand is covered by coniferous forests. The vast majority of the country's plant species are found nowhere else in the world, but a species of North American pine is widespread after being imported to control erosion. Deer and opossum were also introduced, causing widespread damage to native vegetation. A wide range of unique wildlife developed because New Zealand lacked large predators. Flightless birds include the kiwi, moa (now extinct), takahe, and weka.

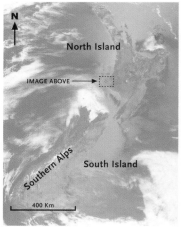

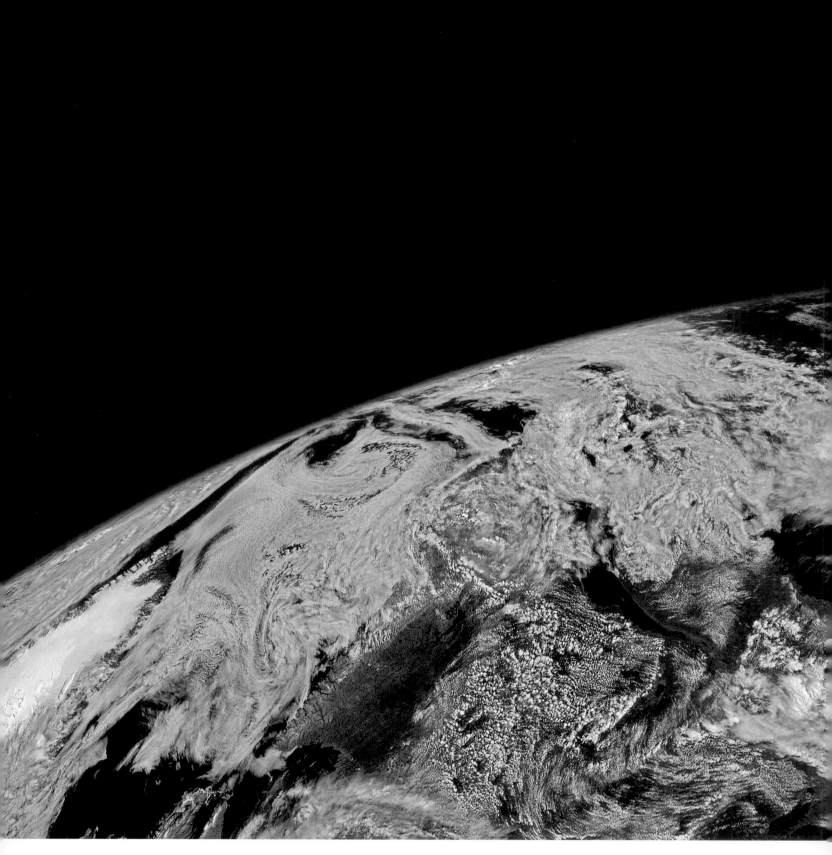

Hurricane on the Move
HURRICANE ALBERTO

Huge weather systems born around the tropics are called hurricanes in North America and typhoons in Asia. Strong winds circulate inward toward an area of low atmospheric pressure at each storm's center. The Earth experiences about 50 to 100 hurricanes and typhoons each year. Although these storms do not last long, they have important effects on global climate because they move trillions of tons of warm, moist air from equatorial areas to higher latitudes.

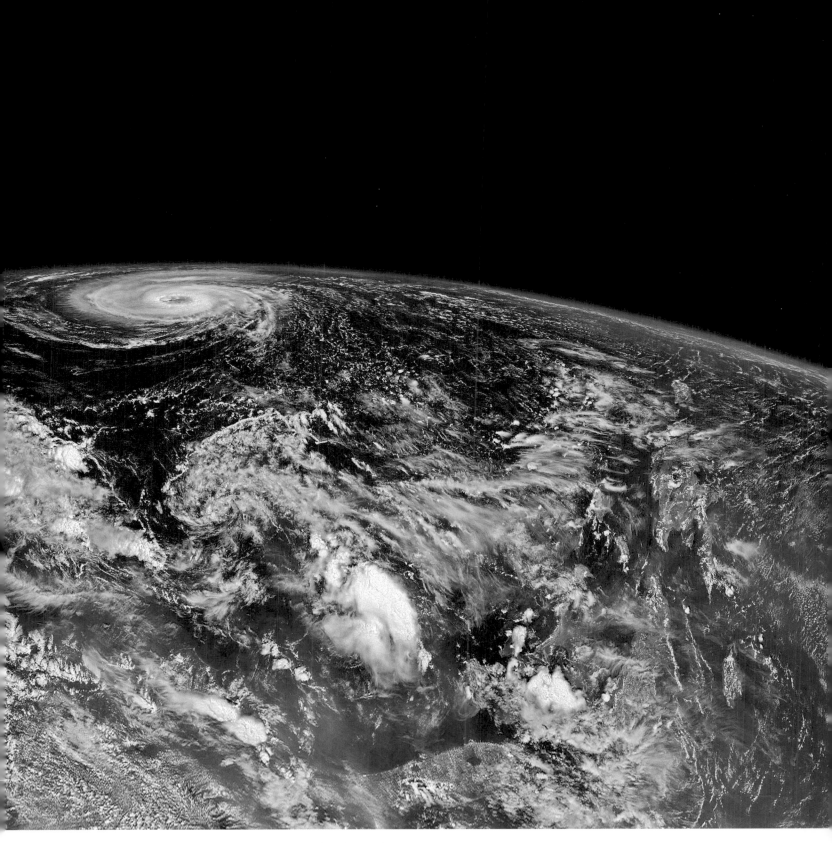

This image shows Hurricane Alberto in the North Atlantic on August 11, 2000. It was collected by the SeaWiFS sensor on the OrbView-2 satellite when Alberto measured hundreds of miles across, reaching its greatest strength east of Bermuda. The satellite was observing this storm obliquely, instead of looking straight down. Alberto lasted about three weeks, making it the longest-living tropical cyclone ever to form in the month of August. It traveled throughout the North Atlantic but never reached land.

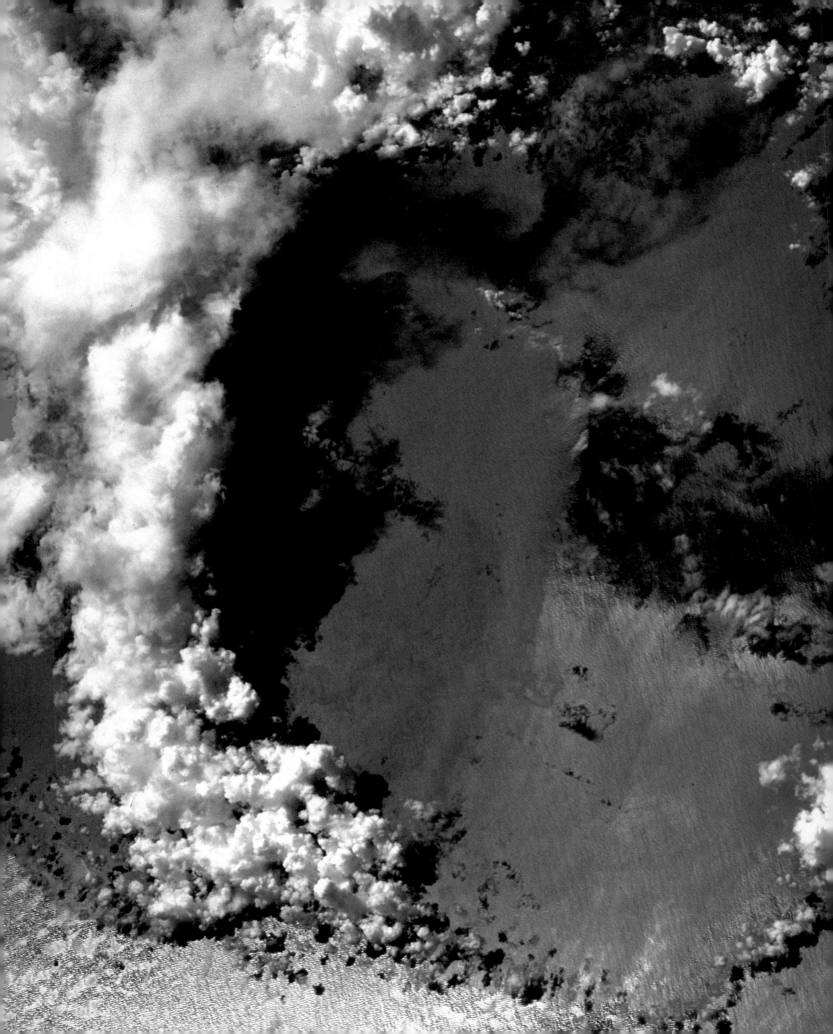

After the Storm

PACIFIC OCEAN (LEFT)

Thunderstorms are created by warm, moist air rising to altitudes of over 12 miles (20 km). As the warm air rises, it cools and water vapor begins to condense and fall as rain. Precipitation can cause downdrafts that slow the rise of more warm air. Eventually the storm runs out of energy and dissipates. This photograph shows the remnants of a thunderstorm near the Micronesian Islands in the Pacific Ocean. The ring of clouds is all that is left of a large thunderstorm. This photograph was taken by Space Shuttle astronauts in October 1994.

Hurricanes

WESTERN HEMISPHERE (ABOVE)

The Earth's atmosphere is key to all life on the planet. Weather patterns are tracked by meteorologists using remote sensing satellites. Satellites in geostationary orbit keep watch over entire hemispheres. This image from the GOES 8 satellite on September 2, 1999, shows continental land masses in green, oceans in blue and clouds in white. Three hurricanes can be seen approaching North America.

Massive Storms

PACIFIC OCEAN (LEFT)

Satellite imagery can be used to understand the evolution of tropical storms as they gain strength and eventually weaken. The image on the left, collected by the MODIS sensor on the Aqua satellite on August 9, 2006, shows three tropical storms swirling over the Pacific Ocean south of Japan. The first to develop was Typhoon Saomai, visible on the lower right. Two smaller storms then moved into the area: Tropical Storm Maria, located on the upper right, and Typhoon Bopha, seen on the left.

Life and Weather

CARIBBEAN SEA (ABOVE)

This view of Cuba and the Caribbean from the MODIS sensor on the Aqua satellite shows atmospheric circulation and life in the oceans.

Cuba is home to a wide range of tropical vegetation, made possible by warm, moist air that arrives from the east. Unfortunately this moist air also brings hurricanes and their destructive force. The eastern end of Cuba receives more rainfall than the rest of the island. Mountains run along most of the length of Cuba, with the highest peaks soaring more than 6,000 feet (1,800 m) on the southeast.

Thousands of native species thrive in Cuba's forests and plains. Palm trees and banana trees grow throughout the island. Much of the native vegetation has been cleared to make way for sugarcane, which has played an important part in the economic development of Cuba.

The Bahamas are visible to the north of Cuba. Around these islands, the light blue shows shallow ocean waters. The darker area is known as the "Tongue of the Ocean," where deeper water is surrounded by shallow reefs.

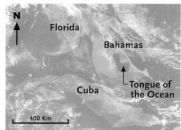

Tropical Convergence

Patterns of atmospheric circulation determine where life exists on the Earth. The most diverse places for life are the Earth's tropical forests. These zones exist near the equator because atmospheric circulation causes greater rainfall there. In these areas the sun heats ocean water. The air is also warmed and subsequently rises. As the air rises, it cools and loses its moisture. This causes a band of intense rainfall all the way around the planet called the Intertropical Convergence Zone, or ITCZ.

At higher latitudes, the Earth experiences warm and cold seasons. Within the ITCZ, the seasons are wet and dry with relatively constant temperatures. However, it can change through time. If the zone moves north or south, it can drastically change rainfall in the affected area, causing drought or flooding.

The ITCZ is visible as the band of bright white clouds across the center of the image. The image was created by combining data from the GOES 11 geostationary weather satellite and landcover data.

Whirlpool

ATLANTIC COAST OF GREENLAND (LEFT)
Cyclones can form if conditions are right.
This Landsat 7 image shows a whirl of
spinning air and water vapor off the coast
of Greenland on May 14, 2001. Blocks of
ice are floating in the ocean water.

Path of Destruction

MARYLAND (ABOVE)
Tornadoes are small cyclones, but they
can cause enormous damage. Tornadoes
form when unstable air in thunderstorms
creates rotating columns of air with
winds that can reach speeds of 300 miles
(480 km) per hour. Tornadoes occur
throughout the world but are especially
common in central North America,
where cold air from the north meets
warm, moist air from the Gulf of Mexico.
More than a thousand tornadoes hit the
United States every year, including about
a dozen very powerful ones.

On April 28, 2002, thunderstorms
swept across Maryland, causing at
least two tornadoes. One of these was
classified as an F4 storm, among the
most damaging. This was the strongest
tornado to hit Maryland and perhaps
the strongest ever recorded in the
eastern United States. It left a path of
destruction over 24 miles (39 km) long.
Three days later, the skies were clear
when the EO-1 satellite passed overhead.
The town of La Plata is visible on the left
side of the image. The storm left a path
of flattened vegetation that reflected
more sunlight, clearly indicating the
tornado's path.

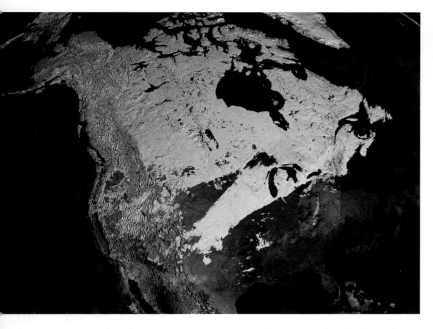

Snow Blankets
the Land

NORTH AMERICA (ABOVE)
CENTRAL UNITED STATES (RIGHT)

Snow falls on almost a quarter of the
Earth's land surface. By cooling the
ground and reflecting sunlight, snow
lowers land temperatures. Data from the
MODIS sensor on the Terra satellite
were used to create the map above
showing the extent of snow cover on
North America. The land surface covered
by snow in February 2002 is shown in
white.

The image on the right shows the
effects of an individual snowstorm. On
December 23 to 25, 2002, a large storm
moved across the central United States.
By the time it reached the eastern United
States, the storm was dumping 5 inches
(12.5 cm) of snow per hour on the
ground. Up to 3 feet (90 cm) of snow
fell, and 19 deaths were blamed on the
storm. On December 25 the MODIS
sensor on the Aqua satellite acquired
this image. White areas show where the
snowstorm crossed the land; the brown
areas are bare ground where no snow
fell. State boundaries have been added
showing the states of Texas, Oklahoma,
Kansas, Missouri and Arkansas.

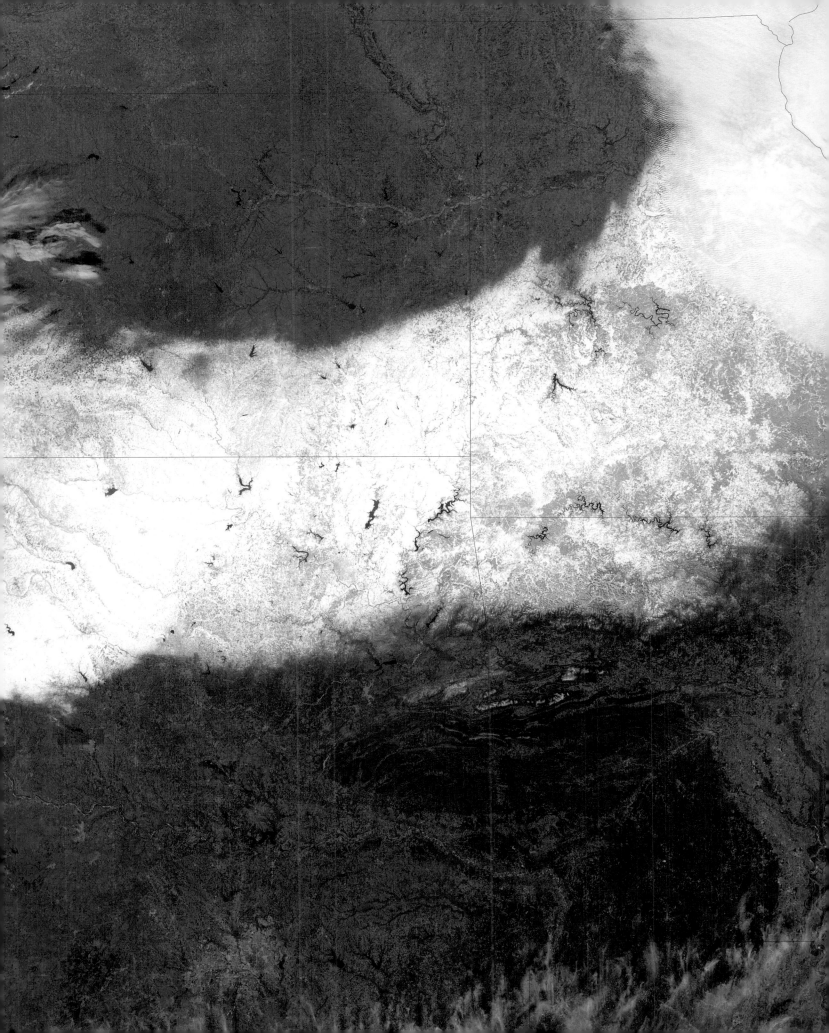

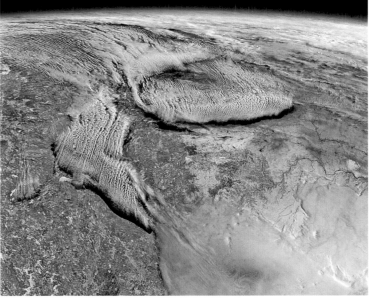

Winter Weather Systems

GREAT LAKES (LEFT)

The image on the left shows the Great Lakes area, which receives a thick blanket of snow every winter. To the north of the Great Lakes, most of Ontario is covered in snow. This image was acquired by the MODIS sensor on the Aqua satellite on December 3, 2002. State and provincial boundaries have been added to the image.

LAKE EFFECT SNOW (ABOVE)

The image above shows how the Great Lakes create their own snowfall, a phenomenon known as the "lake effect." The lakes create their own snowfall when cold air moves south from Canada and picks up water vapor from the lakes. As the air mass flows farther south into the United States the water vapor falls back to Earth as snow. Cities downwind of the Great Lakes often receive huge snowfalls. This image, acquired by the SeaWiFS sensor on the OrbView-2 satellite on January 17, 2002, shows snow thickening over western Michigan. More than 20 inches (50 cm) of snow fell in Kalamazoo, Michigan, in that month.

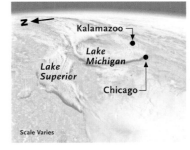

Kalamazoo

Lake Michigan

Lake Superior

Chicago

Scale Varies

Island Wakes

When wind blows clouds around an obstacle such as an island, large swirls are formed in the atmosphere. If the wind blows for a significant time, long strings of vortices, known as vortex streets, are created in the atmosphere.

The central image shows wakes and vortices made by stratocumulus clouds downwind of the Canary Islands in the Atlantic Ocean, about 70 miles (115 km) off the coast of Africa. It was acquired by the MODIS sensor on the Terra satellite in July 2002.

An image from the Landsat 7 satellite (upper left) shows patterns of swirling vortices on June 5, 2000, around the Kuril Islands that stretch between Russia's Kamchatka Peninsula and Japan. An image collected by the MODIS sensor on the Terra satellite in August 2002 (middle left) shows South Georgia Island in the southern Atlantic Ocean. The island is covered in snow and ice, and a wake of stratocumulus clouds stretches to the northeast. The lower-left image shows Attu Island, the westernmost of the Aleutian Islands off Alaska. The Landsat 7 satellite collected this image in July 2000. The MODIS sensor on Terra acquired an image on September 17, 2002 (upper right), showing another set of wakes forming downwind of the Kuril Islands. The lower-right image is a photograph taken by astronauts on the Space Shuttle in November 1994. It shows the Japanese island of Sado surrounded by low-level clouds. The wind is blowing from south to north (toward the lower left), creating an island wake through the cloud layer.

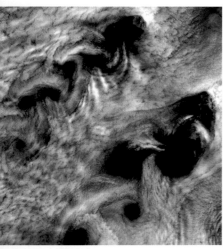

▲ KURIL ISLANDS

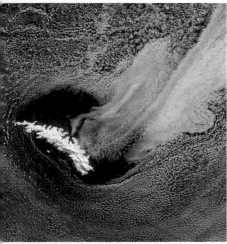

▲ SOUTH GEORGIA ISLAND

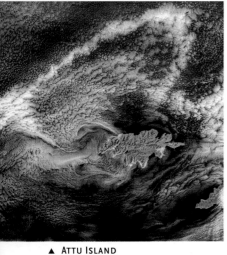

▲ ATTU ISLAND

CANARY ISLANDS ▶

▲ KURIL ISLANDS

▲ SADO ISLAND

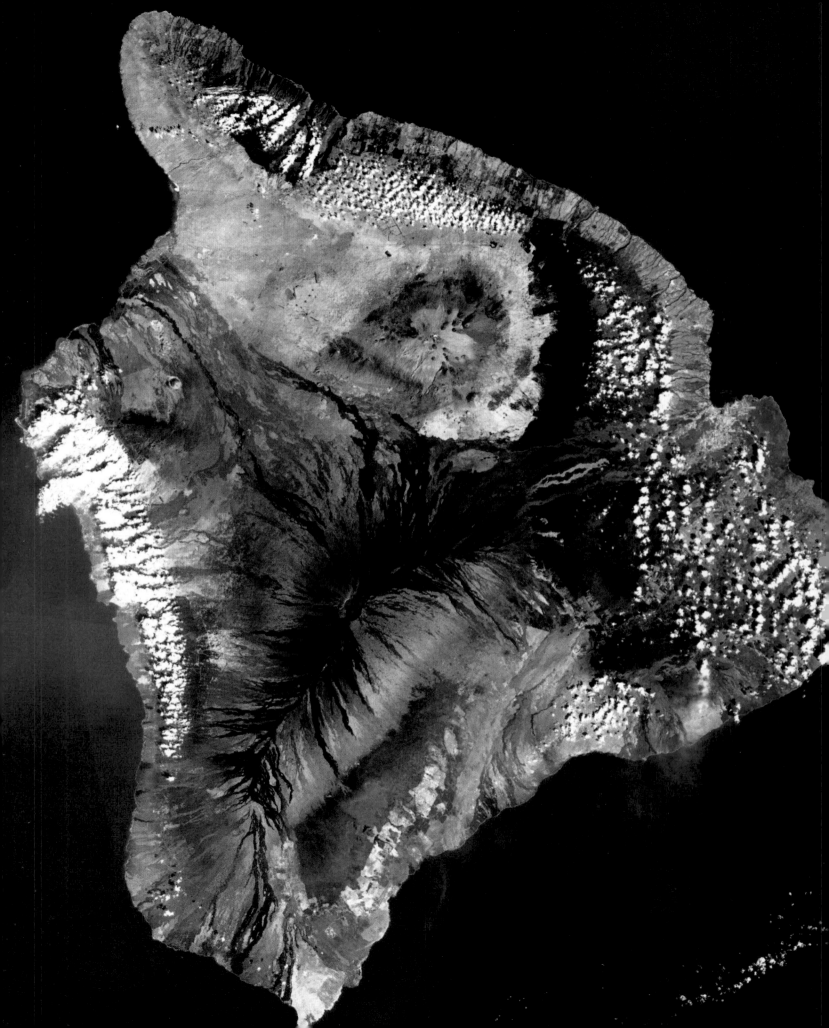

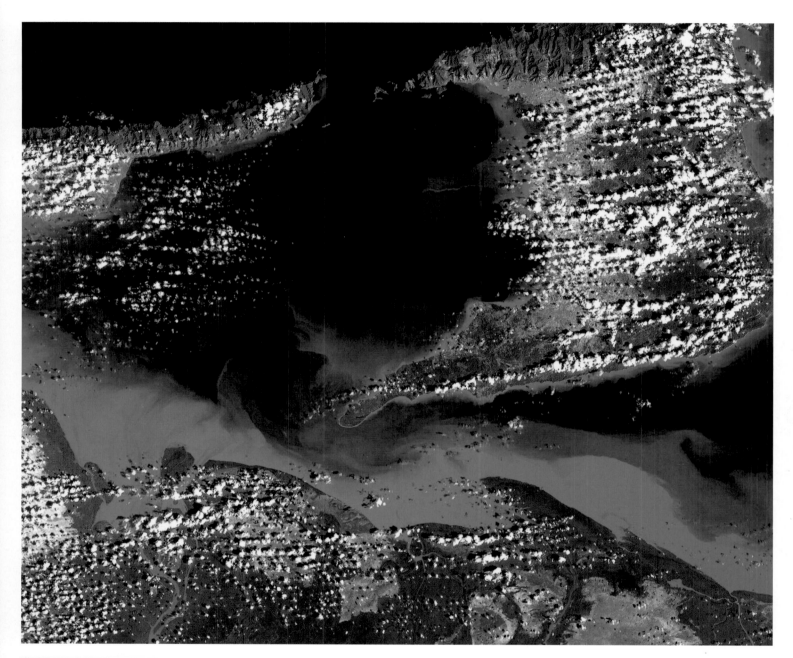

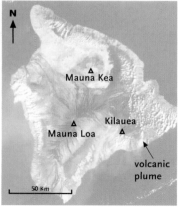

Island Ecosystems

HAWAII (LEFT)

The island of Hawaii, lying in the center of the Pacific Ocean, has great environmental extremes. The dark reds and browns in the image indicate dense vegetation flourishing in warm and wet conditions on Hawaii's eastern coast. There, winds carry clouds onto the land, causing significant precipitation. On the western coast, the drier conditions are ideal for the cultivation of coffee. Much of the island is covered by recent lava flows on which very little vegetation is found. In the center of the island the peaks of Mauna Kea and Mauna Loa are almost 14,000 feet (4,250 m) high. Active volcanoes sit on the island's southeast side, from which a plume of smoke and steam is clearly visible. Since 1983 lava flows have covered 40 square miles (100 sq km). This image was created using Landsat 7 data collected in 2001.

TRINIDAD (ABOVE)

At the eastern end of the Caribbean, the MSS sensor on the Landsat 4 satellite captured an image of Trinidad and part of the Venezuelan mainland on February 4, 1977. The bright red indicates dense vegetation. Sediment can be seen floating in the waters between the island and the mainland. Trinidad has a varied topography. At low altitudes, swampy areas are home to a diverse range of birds. Mountains on the north side of the island reach over 3,000 feet (900 m) above sea level, and there are two waterfalls 300 feet (90 m) high. The tropical climate has created large zones of dense rain forests on the island.

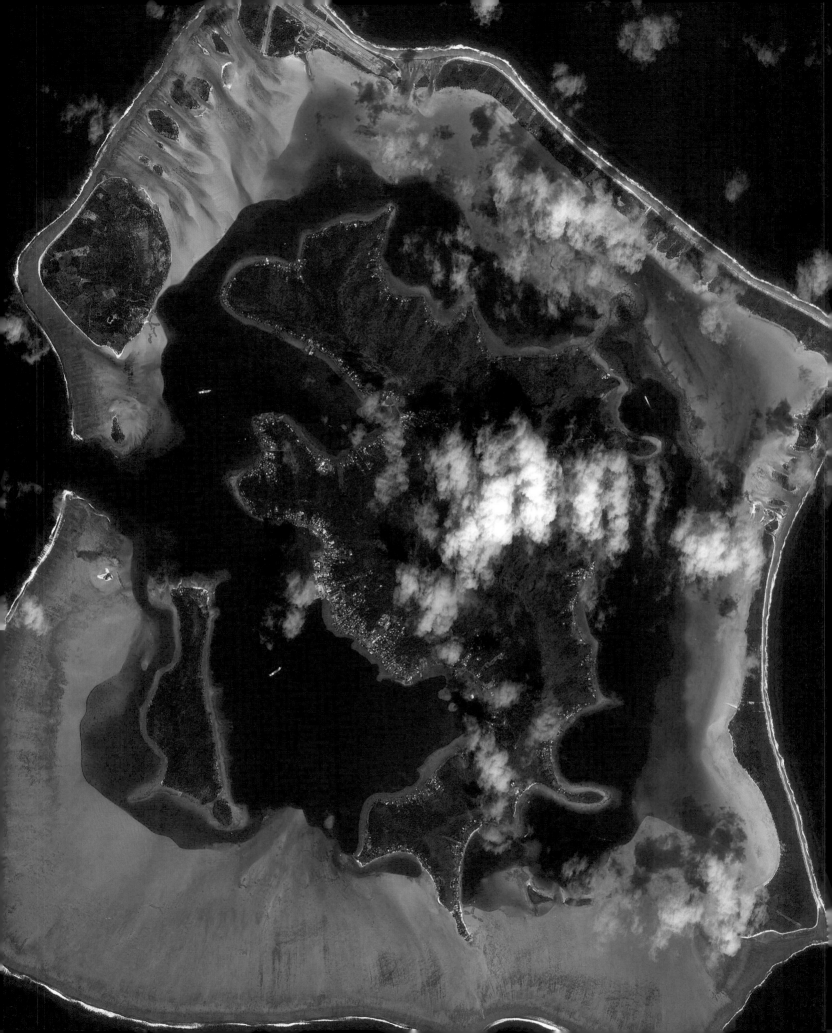

Pacific Islands

BORA-BORA (LEFT)

Many islands in the oceans are surrounded by reefs that are rich in sea life. The image on the left, from the IKONOS satellite, shows Bora-Bora in the Society Islands, part of French Polynesia in the Pacific Ocean and about 150 miles (240 km) northwest of Tahiti. The island was much larger when it was formed by volcanic activity, but millions of years of erosion have resulted in circular landforms surrounding the central island. The reefs protect the interior lagoon, where sea life thrives. The highest peak on Bora-Bora is Mount Otemanu, at 2,380 feet (720 m). Vaitape, the main city, is situated on the west side.

AITUTAKI ATOLL (ABOVE RIGHT)

This IKONOS image shows one of the Cook Islands in the Pacific. The 15 islands in the group are home to about 14,000 people. This image shows the southeast part of Aitutaki Atoll, the second most populous island in the group. Although the Polynesian people of Aitutaki have lived on the atoll for many centuries, the island group is named for James Cook, who visited in the 1770s.

BAKER ISLAND (BELOW RIGHT)

After millions of years of erosion, islands become smaller atolls where just a simple circle of land remains. This image from the IKONOS satellite shows Baker Island, about 1,650 miles (2,700 km) southwest of Hawaii. The yellow color shows reefs under shallow water, and the blue indicates deeper water. An abandoned airstrip crosses the island.

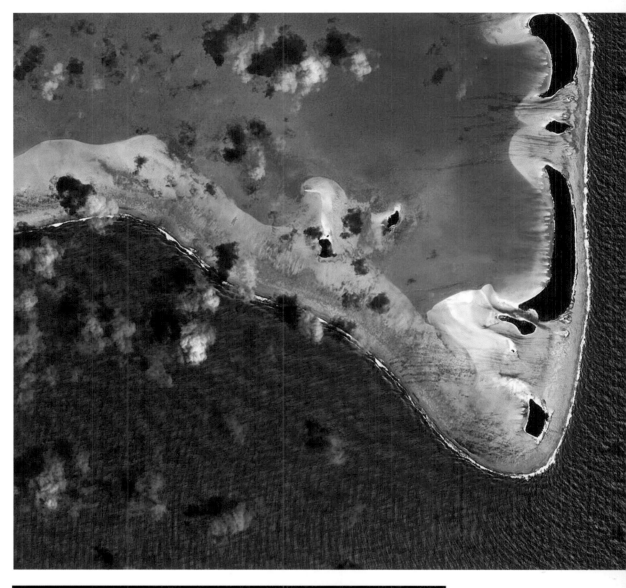

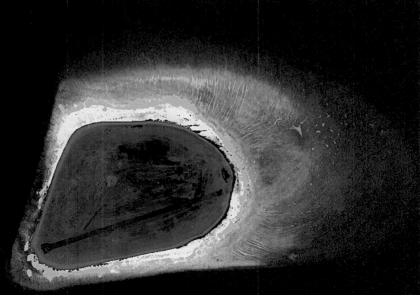

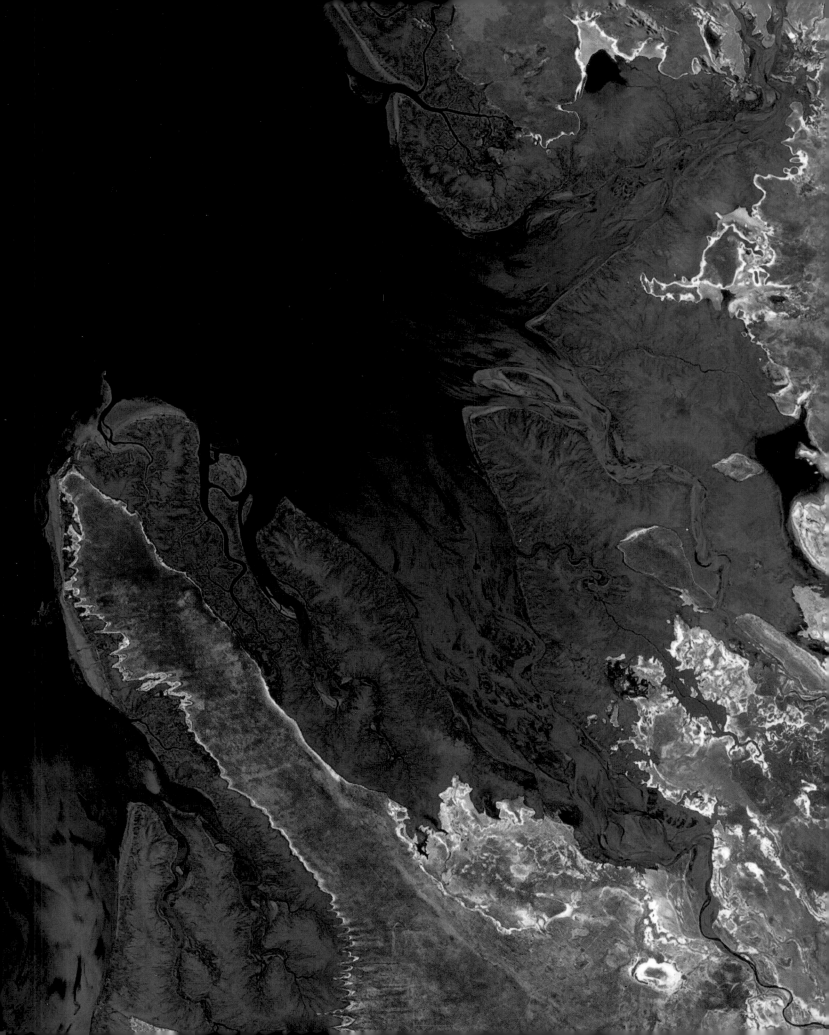

Rivers Meet the Sea

COLLIER BAY, AUSTRALIA (LEFT)

Collected by the Thematic Mapper sensor on the Landsat 5 satellite on March 31, 1990, this image shows the area around Collier Bay in northwestern Australia. Rivers flowing out to sea provide nutrients to organisms in the water just off the coast, allowing plant life to thrive. Green areas off the shoreline indicate areas covered by wetland vegetation, where the land is often flooded.

Largest Reef on Earth

GREAT BARRIER REEF (ABOVE)

The world's largest coral reef, the Great Barrier Reef off the northeastern coast of Queensland, Australia, is made up of almost 3,000 individual coral reefs and coral islands up to 100 miles (160 km) off the coast. The reefs are built of coral — an organism that lives on the sea floor in shallow water. The Great Barrier Reef is more than 1,200 miles (2,000 km) long and runs through warm and very clear water in which 100 foot (30 m) visibility is common.

Rich in diverse ocean life, the Great Barrier Reef is home to worms, crabs, a wide range of fish and hundreds of types of coral. This image was acquired by the MISR sensor on the Terra satellite on August 26, 2000.

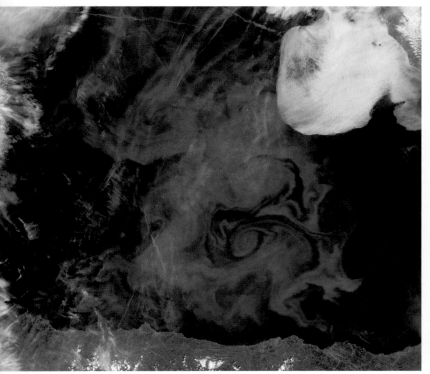

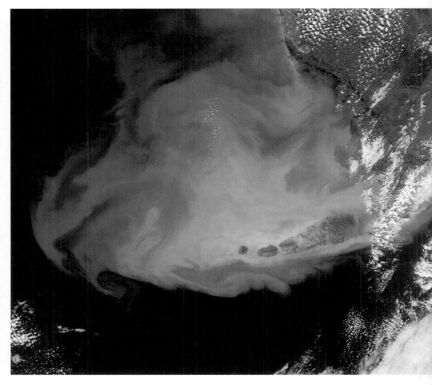

Explosions of Sea Life

CAPRICORN CHANNEL, AUSTRALIA (LEFT)

River currents laden with sediment carry nutrients into the ocean, and cold water welling up from ocean depths brings nutrients to the surface, leading to explosions of sea life called plankton blooms. Phytoplankton usually live about 15–30 feet (5–10 m) below the water surface and is visible from space in high concentrations.

The image on the left shows a plankton bloom that occurred in the Capricorn Channel near the Great Barrier Reef off Australia. The Swain Reefs, at the southern end of the Barrier Reef, can be seen in this image. The plankton forms long filaments that drift into the ocean. Its greenish color is caused by the chlorophyll contained within the plankton. This photograph, which covers an area about 60 miles (100 km) wide, was taken by astronauts on the Space Shuttle in December 1983.

OFF THE COAST OF FRANCE (FAR LEFT)

This image, acquired by the MODIS sensor on the Aqua satellite, shows a plankton bloom off the coast of France on April 30, 2005. Plankton gives a green color to the ocean waters, and tan colors indicate a mixture of plankton and sediment. The swirling pattern is caused by currents circulating in the Bay of Biscay.

GULF OF MEXICO (ABOVE)

Plankton can become mixed in offshore zones to form unusual plumes in the Gulf of Mexico. In the image above, shallow waters are filled with plankton and sediment washed out of the Florida Everglades. This image was collected by the MODIS sensor on the Aqua satellite on October 28, 2005.

Structure of the Land

The planet Earth is made of about six billion trillion tons of rocks and minerals. Deep within the Earth high temperature keeps much of the material in a molten form, slowly churning about under enormous stress and pressure.

Satellite images are widely used to map and help us better understand geological structures such as mountains, volcanoes and faults. Data from remote sensing satellites are also used to detect different types of rock and identify likely features under the surface.

The Earth's Interior

At the center of the Earth is a core made mostly of iron and nickel. Temperatures there probably hover around 8,000°F (4,400°C). The inner core is surrounded by a layer of molten iron and nickel known as the outer core. The outer core's liquid state allows it to circulate around the inner core.

Because of the amount of iron present in the outer core, this moving mixture conducts electricity. As this material circulates it generates a magnetic field big enough to encompass the entire planet. Some planets have huge magnetic fields (Jupiter's is larger than those of all other planets combined), but others have none at all. The Earth's magnetic field protects the surface of the Earth from harmful solar radiation and causes compass needles to point north and south. Some satellites can detect the magnetic field and observe how it interacts with the Sun.

Beyond the outer core are layers known as the lower mantle and upper mantle, which together make up about two-thirds the Earth's mass. The temperature in the lower mantle can reach 6,300°F (3,500°C), while the hottest temperature in the upper mantle is a relatively cool 1,800°F (1,000°C). Although the temperatures are hot enough to melt rock, the high pressure keeps the mantle solid. But even though the mantle is solid rock, heat from the core causes it to revolve slowly in a circular movement, called convection. Although it may take a decade for the mantle to move just one foot, mantle convection slowly allows heat to

Mountains and Dry Valleys

NORTHWESTERN CHINA

This mixture (opposite page) of mountains, sand dunes and rocks on the desert floor lies in northwestern China. This part of China is arid, and a wide range of geologic features are visible. Running across the center of the image is a fault line where earthquakes are common. The white areas are salt deposits laid down long ago, when large lakes dried up leaving a salty residue. In the northern part of the image, large fields of sand dunes can be seen. The mountain range running across the image forms the border between Xinjiang, an autonomous area within China, and the province of Qinghai to the south. The Landsat 5 satellite recorded this image on March 10, 1990.

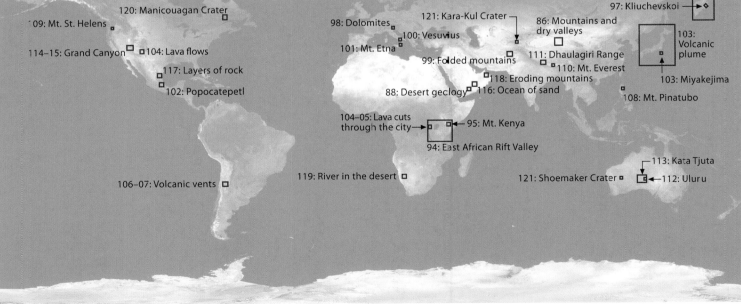

Numbers indicate page locations of the images in this chapter.

are particularly useful for detecting geological differences. Multispectral sensors built to detect these wavelengths enable geologists to make assessments about rock types on the surface. Hyperspectral sensors are even better suited for geological applications. Four or five wavelengths are visible to multispectral sensors, but hyperspectral sensors can detect hundreds of different wavelengths and are therefore better able to distinguish the various minerals present in rocks.

Sometimes it is possible to see beneath the surface and understand recent by about 130,000 meteorites about 100 feet (30 m) across during the last billion years. However, only about 160 impact craters have been identified on the Earth.

The impact of a 100 foot (30 m) wide iron meteorite would leave a crater slightly larger than half a mile wide, leaving a mark visible from space. Much larger impacts also occur. Every 50–100 million years an object about 6 miles (10 km) across may hit the Earth, creating a crater about 60 miles (100 km) across. The amount of energy released in this type of impact is far greater than that contained in all the world's nuclear weapons. An impact of this size would obscure the Sun by throwing huge amounts of dust into the atmosphere and starting massive fires. These events would cause global temperatures to fluctuate wildly, by perhaps as much as 30°F (16°C). This kind of impact may have led to the widespread extinctions of life on the Earth, including the end of the dinosaurs, about 65 million years ago.

Working in the Field

Remote sensing is an important tool for looking at the Earth, but to understand the surface more fully, fieldwork on the ground is needed. Geologists can use a combination of remote sensing and fieldwork to help them understand the Earth. Satellite images help to locate areas of interest, and images from different spacecraft can be compared. A coordinate grid can be added to the images, making it possible to identify interesting features and find them in the field.

To measure accurately the features on the landscape, a variety of surveying techniques are used on the ground. For determining the size and location of objects in a field area, traditional techniques have involved the use of optical instruments such as theodolites. The Global Positioning System (GPS) has revolutionized the way fieldwork is done by allowing scientists to collect a far greater amount of high-precision data in the field.

Remote Sensing of Geology

The view achieved from an orbital satellite allows us to see large areas at one time. This kind of scale is perfect for observing geology. Many remote sensing satellites are equipped with sensors specifically designed to detect differences in surface geology. Mining companies also use satellite images to detect the presence of some minerals and structures likely to contain valuable resources.

The structure of the crystals in a mineral determines how it reflects sunlight and emits infrared energy. Because different rocks contain different minerals, it is possible for satellite sensors to help identify rocks. Certain infrared wavelengths

Science in the Field
Scientists use Global Positioning System (GPS) survey equipment to study the Sabancaya volcano in southern Peru. A view of this volcano from space is seen on page 132.

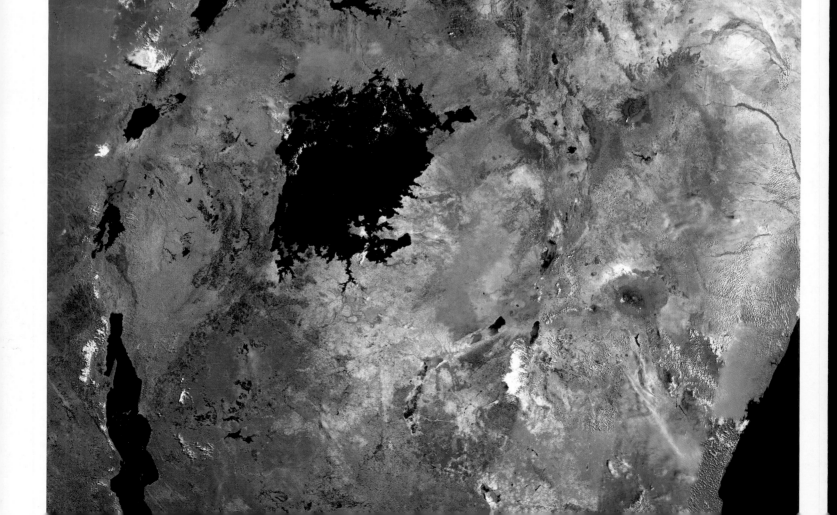

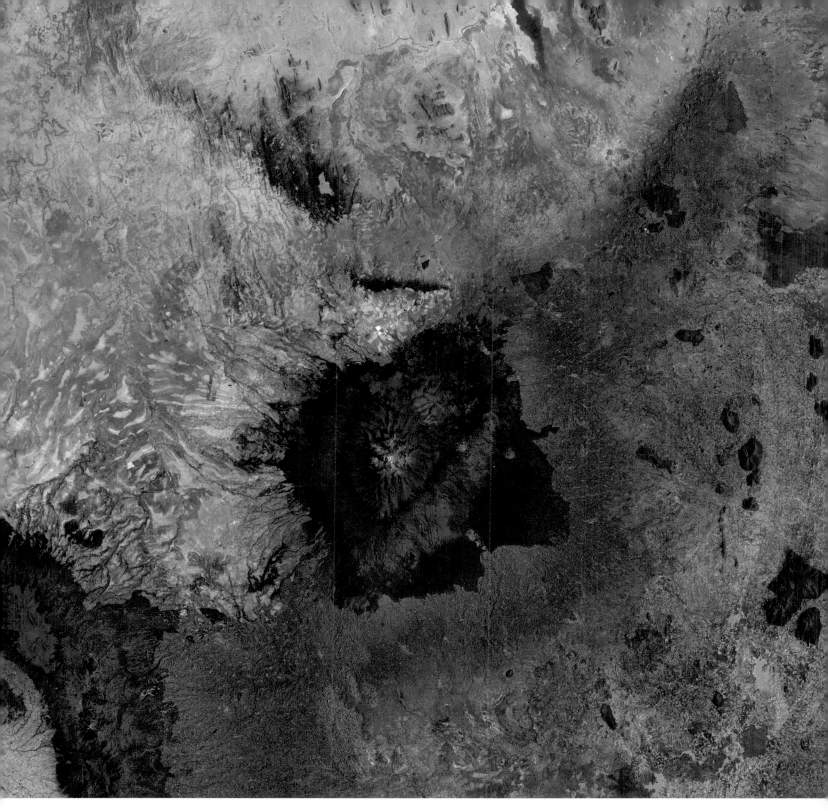

lie to the east of Lake Victoria. This image was produced by combining many different Landsat images.

MOUNT KENYA (ABOVE)

In this close-up view of Mount Kenya, a mountain created by the geological activity around the rift valley, vegetation is shown in red. At the margins of the valley, volcanic peaks such as Mount Kenya (the second-largest mountain in Africa, after Mount Kilimanjaro) rise from the land. Although these peaks are very near to the equator, they remain covered by snow and ice throughout the year. This image, acquired by the Landsat 7 satellite on February 21, 2000, shows the area around Mount Kenya, including much of the Laikipia Plateau to the northwest.

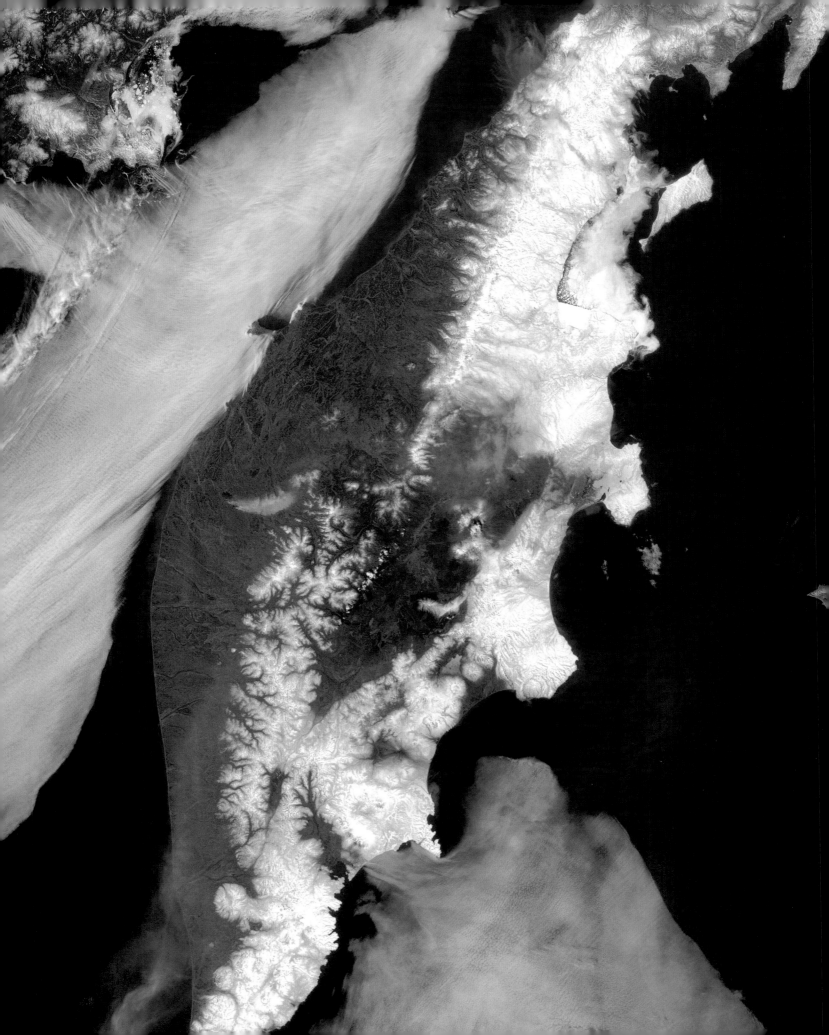

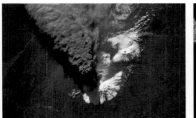

Kamchatka Peninsula

VISIBLE IMAGE (LEFT)

The Pacific Ocean is encircled by a series of volcanoes called the "Ring of Fire." Part of this is the 750 mile (1,250 km) long Kamchatka Peninsula. More than 100 volcanoes can be found here, including at least a dozen that are active. Part of Russia, the Kamchatka Peninsula lies north of Japan. The Central Range Mountains, which run like a spine down the center of the peninsula, contain many active volcanoes. The tallest peaks rise more than 15,000 feet (4,500 m) above sea level. The image on the left, acquired on May 28, 2005 by the MODIS sensor on the Terra satellite, shows the mountains covered in snow. Plumes of smoke from forest fires and ash from volcanic eruptions are visible.

TOPOGRAPHY (RIGHT)

The image on the right shows the Kamchatka Peninsula as it would look if we could strip away the snow and ice to reveal the land. The central mountains are older volcanic peaks. A younger and more active chain of volcanoes follows the Pacific coast. Green indicates low elevations, white high elevations, and brown in between. This image was produced using data from the Space Shuttle Radar Topography Mission in February 2000.

VOLCANIC ERUPTION (ABOVE)

The photograph above, taken by Space Shuttle astronauts on September 30, 1994, shows an eruption of Kliuchevskoi, the largest volcano on the peninsula.

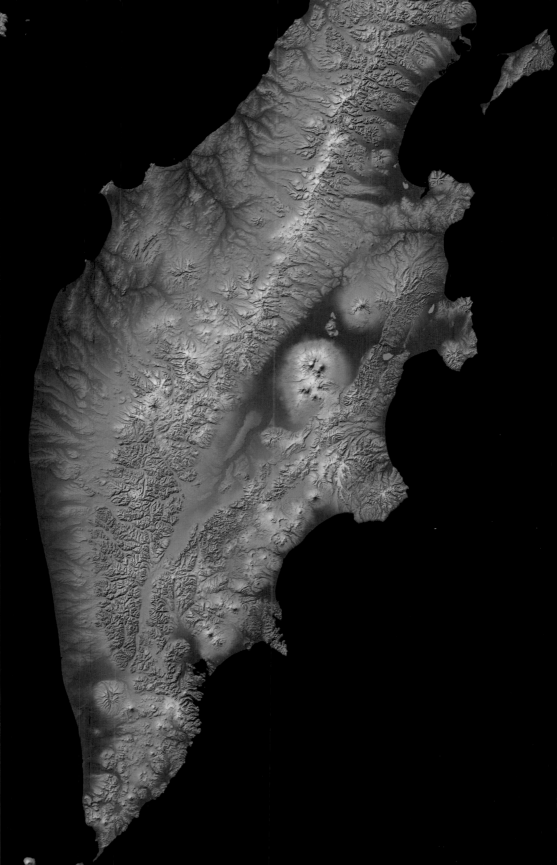

Mountain Building

DOLOMITE MOUNTAINS, ITALY (ABOVE)
This high-resolution view shows Italy's Dolomite Mountains. Part of the Alps that stretch across Europe, the Dolomites lie north of Venice. The Alps were created by continental drift pushing Italy against the rest of Europe. The distinctive light-colored limestone clearly visible here was created long ago on the sea floor. Geological forces pushed the limestone to the top of the mountain range, where millions more years of erosion carved sharp pinnacles that reach thousands of feet into the sky. The IKONOS satellite collected this image on October 25, 2000.

FOLDED MOUNTAINS, AFGHANISTAN (RIGHT)
The rocks that make up our continental plates are slowly moved and bent by geological forces so powerful that, over time, the ground can actually be folded — creating hills and mountains. This image from the Landsat 5 satellite, acquired on October 25, 1988, shows folded and eroded mountains in southeastern Afghanistan on the border with Pakistan. Erosion has worn away the mountains, exposing layers of rock and making them visible from orbit. These mountains lie at high altitudes; the entire area is over 6,600 feet (2,000 m) above sea level. Tributaries of the Kundar River can be seen in the lower-right corner of the image.

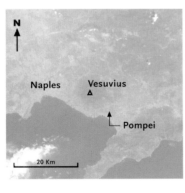

Living with Volcanoes

VESUVIUS (LEFT)

Naples, Italy, is a thriving port city located right next to a large volcano. The volcano, Vesuvius, is famous because of an eruption in 79 AD that destroyed the Roman city of Pompeii, just to the southeast. The volcano is still active, and geologists have measured the presence of molten magma about 25,000 feet (7,600 m) below the surface.

The last eruption of Vesuvius occurred in 1944 and killed about 20 people. Today the Naples metropolitan area reaches to the flanks of the volcano, and 600,000 people live nearby. The size of the surrounding population makes Vesuvius one of the most dangerous volcanoes in the world. The Italian government has been trying to convince residents to move away from the danger, offering payments to those who choose to move from the 18 towns nearest the volcano. Vegetation appears red in this image from the SPOT 5 satellite.

MOUNT ETNA (ABOVE)

Mount Etna, on the island of Sicily, is the tallest volcano in Europe at about 11,000 feet (3,300 m), and has been active for thousands of years. On October 27, 2002, it began to erupt, and at the time was the most active volcano in the world. In this image, a large dark cloud of smoke pours out of the volcano. The plumes of lighter smoke on the left are emissions of gas from vents on the volcano's flanks. This photograph was taken by an astronaut on the International Space Station using a hand-held digital camera.

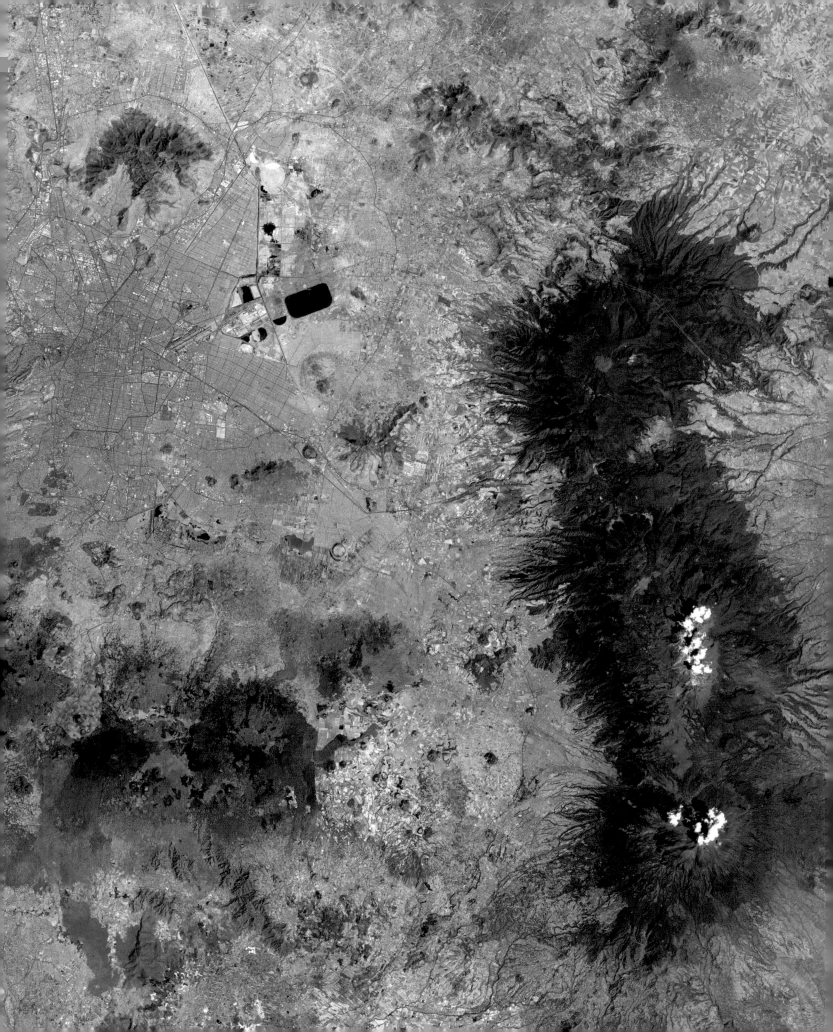

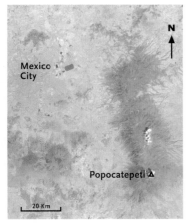

Living with Volcanoes

POPOCATEPETL (LEFT)

About 40 miles (65 km) southeast of Mexico City is a volcano called Popocatepetl, nicknamed Popo, with a summit 17,930 feet (5,465 m) above sea level. Popo, located in the lower right of this image, is covered by small clouds. Under the cloud cover it has a gray appearance because of volcanic ash, in contrast to the snow-covered mountain to the north. The metropolitan area of Mexico City, with 20 million people, is the light gray region in the upper left.

Popo experienced a large eruption 1,200 years ago. Lava from that eruption filled in valleys and blanketed the surrounding area. There have been occasional eruptions in the centuries since, including one on December 18, 2000. During that eruption people were evacuated to avoid the ash, smoke and rocks that were thrown into the air. This image was collected by the Landsat 7 satellite on March 21, 2000.

MIYAKEJIMA (ABOVE)

The collision of continental plates created the islands of Japan, making them one of the most geologically active areas in the world. Earthquakes are common throughout the islands, and active volcanoes are a part of the landscape. More than 60 volcanoes have erupted in Japan during historical times.

The image above shows Miyakejima, one of Japan's Izu islands, about 100 miles (160 km) south of Tokyo. Mount Oyama, an active volcano, dominates the small island. Gases vented from the volcano create white clouds. The green colors show vegetation cover. Gray indicates areas covered by volcanic ash. A large airport is visible on the right side of the island. This image was acquired by the QuickBird satellite on March 14, 2002.

VOLCANIC PLUME (ABOVE)

The plume of volcanic smoke visible in the center comes from eruptions that began at Mount Oyama in July 2000. The smoke hangs over the main Japanese island of Honshu. This image was collected by the MODIS sensor on the Terra satellite on August 29, 2000. Three days later, authorities decided to evacuate Miyakejima.

Lava Flows

NEW MEXICO (ABOVE)

This Landsat 7 image, acquired on April 14, 2000, shows the McCartys Lava Flow in western New Mexico. Red areas indicate vegetation reflecting near-infrared wavelengths; areas covered by lava are black. The flows are almost 30 miles (50 km) long, which is rare on Earth but more common on Venus and Mars. The most recent lava flows here are only about 3,000 years old, making it virtually certain that people were present to witness the lava flow. The Zuni-Acoma trail, connecting two ancient towns, crosses the lava. Acoma was settled by 1000 AD, making it one of the oldest continually inhabited places in the United States.

Lava Cuts through the City

VIRUNGA MOUNTAINS (LEFT)
GOMA (RIGHT)

The Virunga Mountains lie on the border between Rwanda and Congo near Lake Kivu in east central Africa. These volcanic peaks, part of the African Rift Valley, rise about 10,000–13,000 feet (3,000–4,000 m) above sea level. Nyiragongo and Nyamuragira volcanoes lie within Congo's Virunga National Park. Goma, Congo, a city of about 500,000 people, lies on the shores of Lake Kivu.

On January 17, 2002, Nyiragongo erupted, sending lava into Goma. A large lava flow cut through the western part of the city, destroying more than 10,000 homes and killing about 100 people. Data from the Landsat 7 satellite and the ASTER sensor on the Terra satellite were combined to produce the image on the left. Thermal-infrared data collected by ASTER on January 28 were used to detect the heat of the lava. This area is indicated in red. The rest of the image was acquired by Landsat 7 on December 11, 2001. The image on the right was acquired by the EROS A satellite after the lava cooled and hardened. It shows lava covering the runway at the Goma airport and cutting through the city.

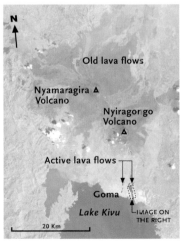

Volcanic Vents
ANDES MOUNTAINS

The Andes Mountains between Chile and Argentina, which contain some of the world's tallest peaks, are home to vast zones of volcanic activity. In some areas ash erupted to form small cone-shaped piles on the landscape. The area shown here is part of the Atacama Province of Chile. It is near the Catamarca Province of Argentina, and is dominated by small dormant volcanoes. Volcanic cones are spread throughout the image. Volcanic activity here has been going on for millions of years, resulting in lava about 2 miles (3.2 km) thick. The different colors indicate the different minerals contained within the rocks. This image was collected by the Landsat 7 satellite on May 15, 1999.

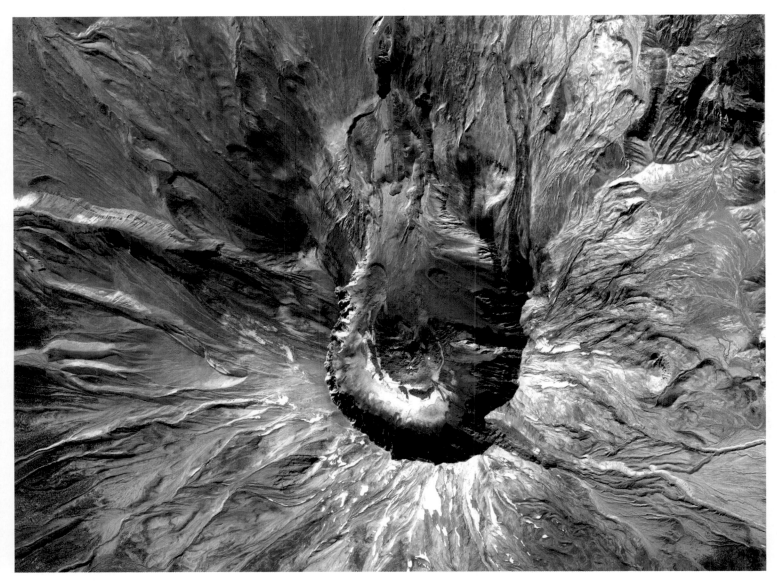

After the Eruption

MOUNT PINATUBO (LEFT)

When a volcano erupts, the surrounding environment is scoured by heated gases and covered by lava or ash. It takes many years before erosion softens the landscape and plant and animal life return.

Mount Pinatubo is located in the Philippines, about 55 miles (90 km) north of Manila. The volcano erupted in June 1991, in what may have been the largest volcanic eruption in a century. The eruption reduced the elevation of the mountain from 5,840 feet (1,780 m) to 4,870 feet (1,485 m). More than 700 people died, and about 100,000 people lost their homes. Volcanic ash mixed with rain subsequently created huge mudslides down the sides of the volcano. A smaller eruption in August 1992 was powerful enough to do serious damage and cause 72 more deaths. Following the eruptions, the environment has been recovering and has undergone changes. The force of the 1992 eruption left a caldera at the summit of the mountain, which, in the years since, has filled with water to form a lake. The IKONOS satellite acquired this image on March 6, 2001.

MOUNT ST. HELENS (ABOVE)

Mount St. Helens is a volcano in the Cascade Range in Washington State. It has erupted twice in historic times, once in 1857 and again on May 18, 1980. The entire top of the mountain was removed by the force of the blast, decreasing the peak's altitude by 1,300 feet (400 m). A cloud of ash rose over 12 miles (19 km) into the atmosphere and spread hundreds of miles away. Mudflows ran downhill for 17 miles (27 km). About 60 people died. Completely destroyed were 10 million trees within 70 square miles (180 sq km) of dense forests surrounding the volcano.

This image from the IKONOS satellite shows the peak of Mount St. Helens on September 13, 2003. More than 20 years after the eruption, the summit area still resembles a moonscape. However, forests have made a comeback at lower elevations, visible in the lower corners of the image. A new dome has been created in the center of the summit crater by hot magma slowly being pushed upward.

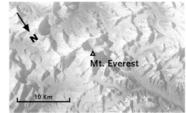

Earth's Tallest Mountains

MOUNT EVEREST (LEFT)

The image on the left shows Mount Everest, known as Chomolungma in Tibetan. At 29,035 feet (8,850 m) above sea level, it is the tallest mountain in the world. The peak of Everest rises above two-thirds of the Earth's atmosphere, and the low oxygen levels can be deadly to those who climb the mountain. The U-shaped valley just south of the peak is the path taken by the first climbers to conquer Everest in 1953, and it remains the most popular route to the top. This photograph was taken by Space Shuttle astronauts in November 1994.

In the mid-1800s, the British Survey of India identified Mount Everest as the tallest mountain in the world. The survey used theodolites, instruments that calculate altitudes of distant objects by measuring their angle above the horizon. Surveyors, observing the mountains from 200 kilometers (120 miles) away, were required to correct for the effects of the Earth's curvature, refraction of the atmosphere, and the gravitational pull of large mountains. Despite the difficulties, the altitude measured for Everest was within 10 meters (33 feet) of the figure determined later through the use of advanced Global Positioning System (GPS) devices. The peak was named for George Everest, who was superintendent of the Survey of India until 1843.

DHAULAGIRI RANGE, NEPAL (ABOVE)

The Himalayan Mountains form a long line of peaks separating Tibet from Nepal, India and Pakistan. About 60 million years ago the Himalayan Mountains began to rise where the Indian subcontinent and Asia collided. The mountains are still growing a few millimeters each year. Glacial ice has eroded them into the shape we see today.

The photograph above, taken from the International Space Station, shows the Dhaulagiri Range in Nepal. The tallest peak in the range is Mount Dhaulagiri, at 26,794 feet (8,167 m) the seventh-highest mountain in the world. The valley in the foreground, at an altitude of about 17,000 feet (5,100 m), contains the most distant sources of the Brahmaputra River, which flows over 1,500 miles (2,500 km) to the Indian Ocean.

Monoliths in the Australian Desert

ULURU (LEFT)

The image on the left shows Uluru, also called Ayers Rock, in Australia's Northern Territory. Uluru stands 1,148 feet (348 m) high, with a circumference of 5.6 miles (9 km), and covers about 1.275 square miles (3.3 sq km). The rock is a monumental geological feature, the largest of its kind. Because the sandstone of Uluru is more resistant to erosion than is the surrounding rock, it stands alone — surrounded by flat arid plains formed after erosion carved away the softer rock.

Its sandstone has a reddish hue, which is especially bright at sunrise and sunset. It has long been a sacred site for the Anangu people, and is now a popular tourist attraction. This high-resolution image, acquired by the QuickBird satellite on February 4, 2002, shows small features on the surface of the sandstone, which were created by erosion.

ULURU AND KATA TJUTA (ABOVE)

The image above shows the area surrounding Uluru. Visible to the west is Kata Tjuta, also known as the Olgas, located about 275 miles (460 km) from the city of Alice Springs. The government of Australia and the Aboriginal community administer the Uluru–Kata Tjuta National Park. Kata Tjuta, which means "many heads," consists of dozens of tall dome-shaped features made of conglomerate rocks. This false color image was acquired by the Landsat 7 satellite on July 12, 1999.

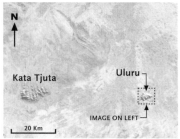

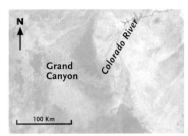

Eroding the Land

THE GRAND CANYON

The Grand Canyon, as seen in this image from the Landsat 7 satellite, stretches about 280 miles (450 km) across northern Arizona. At 30 miles (50 km) wide and 6,000 feet (1,800 m) deep in places, it is the largest canyon on Earth. The Colorado River, running through the canyon, has immense power to erode rocks. Before it was dammed, it moved 500,000 tons of eroded rock through the canyon every day.

The formation of the Grand Canyon began about 6 million years ago. As the land was uplifted by forces deep beneath the Earth's surface, the river continued to flow. While the river cut into the bottom of the canyon, the dry climate limited erosion of the canyon's sides — leaving the steep walls we see today. The oldest rocks, at the bottom of the canyon, are more than two billion years old. Younger rocks lie in the layers above.

The Colorado River supplies more water to people than any other river in the world. Downstream from the Grand Canyon, the cities and farms in Arizona and California use almost all the water flowing through the river, allowing very little to reach the ocean.

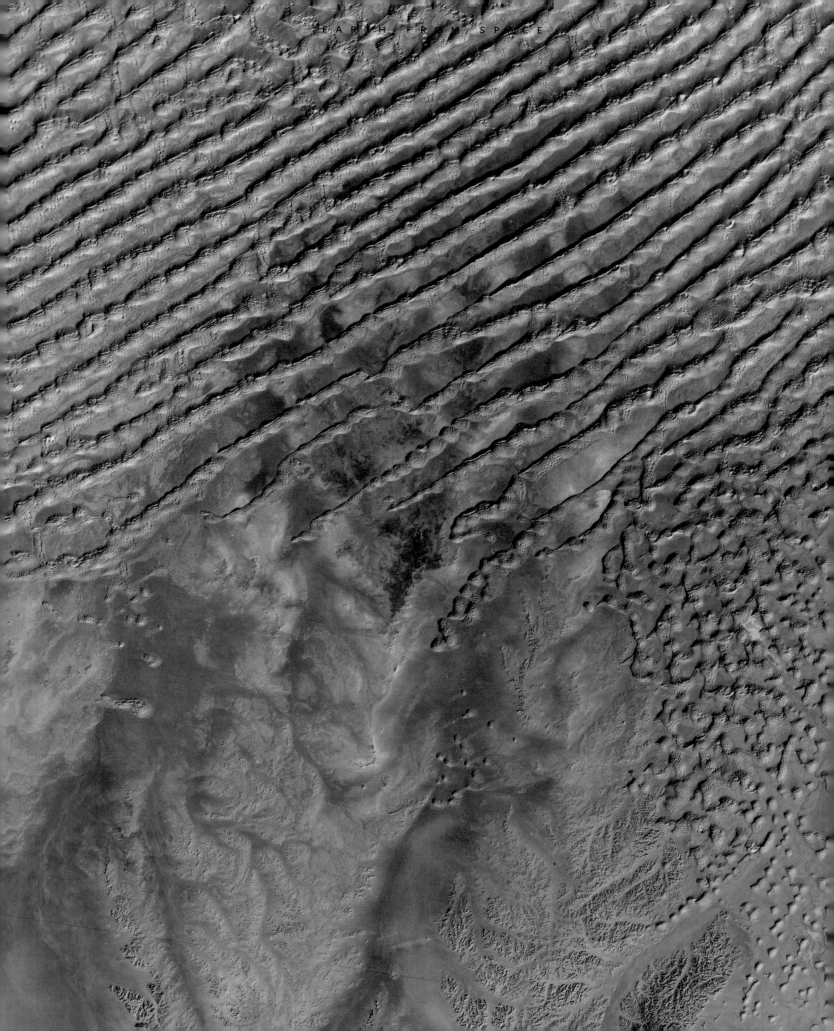

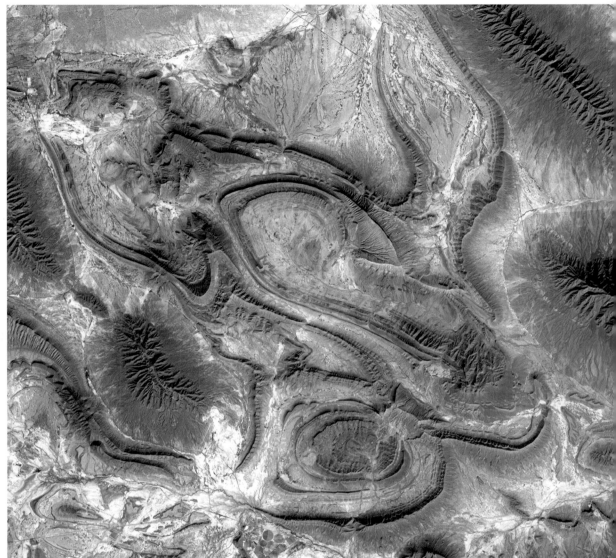

Ocean of Sand

SAND DUNES, YEMEN (LEFT)

This image shows a field of sand dunes in Yemen, near the border with Saudi Arabia. The international border between the two countries has not been surveyed or marked on the ground. The sands visible here are at the southern edge of the Rub' al Khālī (the "empty quarter"), a huge expanse of desert sands. This area is virtually uninhabited.

A rocky desert surface is seen to the south along with drainage features. The Wadi Armah river flows intermittently from Yemen to the north, drying up as it reaches the sand. Infrared wavelengths were used to produce the bright colors of this image. Blue indicates darker and more rocky composition of the land and contrasts sharply with the highly reflective sand dunes shown in yellow. This image is a combination of data recorded by the SPOT 1 and Landsat 5 satellites.

Layers of Rock

MOUNTAINS, MEXICO (ABOVE)

This image shows the Sierra Madre Oriental Mountains in northern Mexico, which form part of the boundary of the central Mexican plateau. These mountains are made of folded sedimentary rocks about 100 million years old that have been uplifted and eroded to reveal the structure of the underlying sedimentary rocks. Little grows in this arid area. The only dense vegetation visible is made up of several irrigated agricultural fields to the south. This image was collected by the Landsat 7 satellite on November 28, 1999.

Eroding Mountains

NORTHERN OMAN (ABOVE)

This image shows the Al Hajar Mountains in northern Oman. These peaks run parallel to the coast of the Gulf of Oman and rise to elevations above 4,000 feet (1,200 m). The lack of vegetation in this dry climate makes the geology plainly visible from space. Millions of years of erosion have left the mountains surrounded by loose, rocky terrain, which appears a purplish color in this image. These areas reflect strongly in visible wavelengths. The green color indicates infrared wavelengths reflected by bare rocks higher in the mountains.

The deserts and mountains visible in this image isolated Oman from the rest of the Arabian Peninsula, allowing cities on the coast to develop and flourish. In the 1400s mariners from Oman were trading goods along the coast of Africa and traveled as far as China. This image from the Landsat 7 satellite shows an area about 120 miles (200 km) southwest of Muscat, Oman's capital and largest city.

River in the Desert

NAMIBIA (RIGHT)

The Namib Desert in Namibia is one of the driest places on Earth. This image shows the Tsauchab River, one of the few sources of water in the desert. This intermittent river has created a clearing through sand dunes, called the Sossos Vlei. Water rarely flows as far as the left side of this image, where the desert finally claims the river. The bright yellow areas are covered by sand dunes up to 980 feet (300 m) high. The long north-south lines of dunes often stretch for many miles. This image was collected by the Landsat 7 satellite on August 12, 2000.

Rivers flowing west into the Namib Desert all end in a vlei, or mud flat, next to desert sands. The occasional presence of water prevents sand from covering the stream channel. Some of the Namib Desert's streams continue to flow under the surface and have been tapped by pipelines to supply water to cities on the Namibian coast.

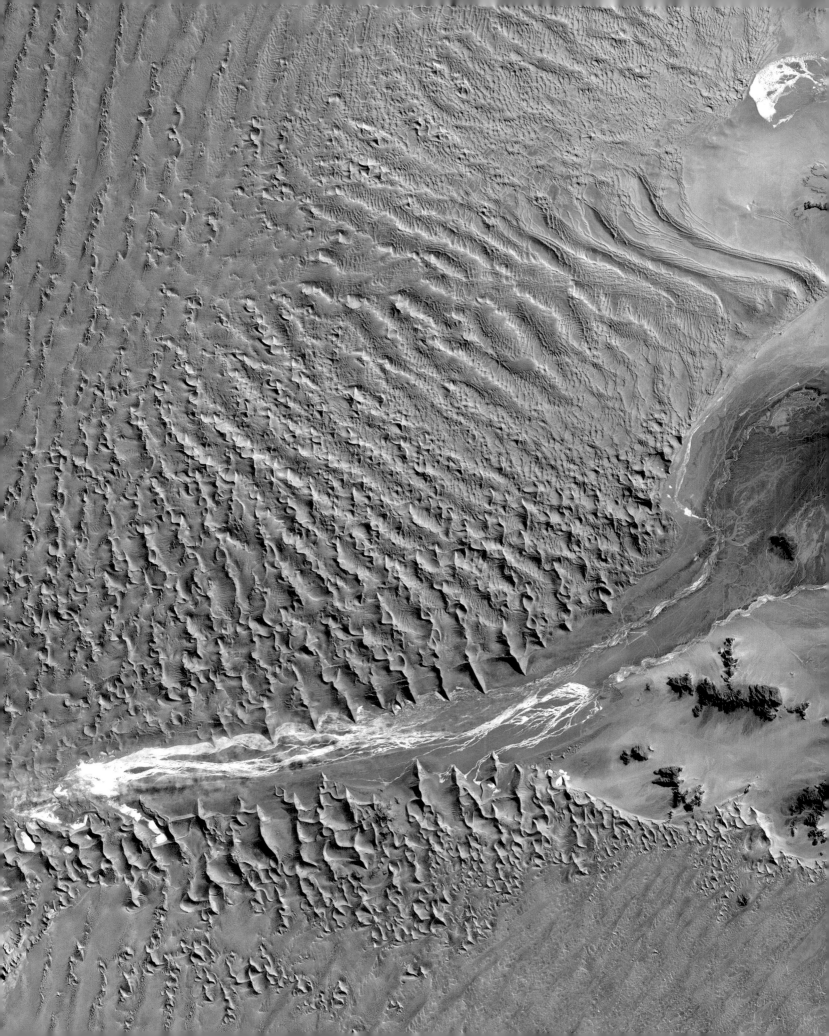

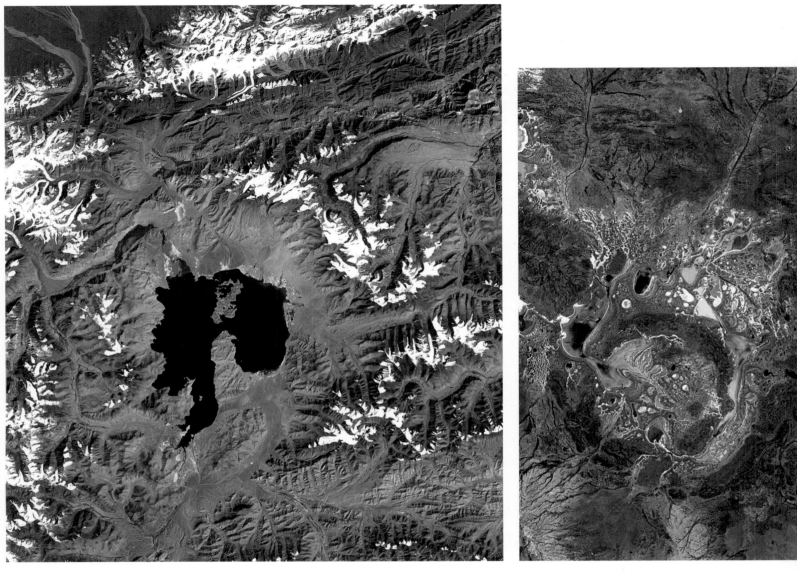

Impact Craters

MANICOUAGAN (LEFT)

Objects from space that strike the Earth's surface leave behind craters in a wide range of forms. Manicouagan, one of the largest craters, is shown here. About 214 million years ago a meteorite about 3 miles (5 km) wide slammed into what is now Quebec. In the center of the crater geologists have found shattered rocks, called breccias, which indicate a massive impact. Huge amounts of dust kicked up by the impact may have led to the extinction of many species of life. Manicouagan Crater may have been as large as 60 miles (100 km) across before it was eroded by glaciers. The crater's edges are still plainly visible, but the peak that once may have stood

at its center has eroded away. Lake Manicouagan fills the depression around the crater. This image was collected by the Landsat 7 satellite on June 1, 2001.

KARA-KUL (ABOVE LEFT)

The Kara-Kul impact crater, created about 25 million years ago, is in the rugged Pamir Mountains in eastern Tajikistan, at an elevation of about 20,000 feet (6,000 m) above sea level. Because there are no roads through this isolated area, the crater was unknown until it was identified in satellite images such as this one. A lake fills the impact crater. Kara-Kul, much smaller and younger than Manicouagan, has a diameter of about 28 miles (45 km).

The island in the lake is a central peak, a common feature of impact craters. This image was acquired by the Landsat 7 satellite on September 28, 2001.

SHOEMAKER (ABOVE RIGHT)

Shoemaker Crater was formed by an impact in western Australia about 1.7 billion years ago, which makes it the oldest known crater on the continent. It is 18 miles (30 km) across and lies in the desert about 100 miles (160 km) west of Lake Carnegie. The outer part of the crater is made up of sedimentary rocks over 2 billion years old, while the inner crater is made of granite. This part of western Australia is an arid desert, causing the lakes within the crater to evaporate and leave behind

salt deposits. Originally known as Teague Crater, it was later renamed for geologist Eugene M. Shoemaker, who played an important role in understanding impact craters. This image was acquired by the Landsat 7 satellite on May 5, 2000.

Water and Ice

Water is the most powerful force changing the Earth's surface. In its liquid and solid form, water moves across the land surface and modifies it in many ways. Rivers flow through the landscapes, eroding geological layers. Ice forms glaciers that slowly move downhill and carve huge channels through mountains all over the world.

The existence of water is one of the things that makes our planet special. We know of only two other places in the solar system where liquid water may exist. On Europa, a moon of Jupiter, water may exist under an icy surface. Water played a part in the geological history of Mars, where small amounts may still exist in places. However, Earth is the only planet where a large amount of liquid water is easy to see today.

Water on the Land

Fresh water falls as rain and snow and collects in low-lying areas to form lakes and ponds. Although fresh water accounts for only one-half of one percent of the water on Earth, it is necessary for humans and most other life on the land. Water makes life possible, and it is present in a great range of places all across the world. Because of the importance of water, locating its sources has always played an important part in the building of human societies. Different populations often compete for water resources when it is scarce.

Whether in solid or liquid form, water can drastically change the appearance of the land. Rivers flow across the surface and erode rocks. Layers of ice form glaciers that shape mountaintops. The effects of water on the Earth's surface are easy to see from an orbital perspective.

When water moves it has the power to wear away the tallest mountains. The

Flowing Ice

LAMBERT GLACIER, ANTARCTICA
For millions of years the continent of Antarctica (opposite page) has been almost completely covered in sheets of ice thousands of feet thick. Glaciers form in the high interior of Antarctica and slowly flow downhill to the ocean.

This image shows the Lambert Glacier, the largest glacier on Earth. Over 200 miles (320 km) long, it twists and turns downhill, dropping about 1,300 feet (400 m) at the point shown here. This is also among the Earth's fastest glaciers, moving more than 4,000 feet (1,200 m) every year. In the lower right of the image is the Amery Ice Shelf, where the ice is floating on the ocean. The Landsat 7 satellite acquired this image on December 2, 2000.

Flowing River

PARANÁ RIVER

As rivers flow across the landscape they modify the land surface. The Paraná River is the second-largest river in South America, after the Amazon. Its tributaries reach throughout southern Brazil, as far as São Paulo near the Atlantic Ocean and to Brasília in the interior. In northern Argentina, the upper Paraná widens into a broad river. This image shows the point at which the Paraná joins the Uruguay River to form the Río de la Plata. The Uruguay River flows from north to south on the right side of the image. The Paraná flows from the west, widening into a broad delta surrounded by forests and marshes. The bright colored lines cutting across the marshes indicate former courses of the river. The city of Buenos Aires is just off the image to the south. The Landsat 7 satellite acquired this image on May 26, 2000.

rocky surface of the Earth does not remain as it is for long. Forces act all across the Earth to slowly wear down high peaks and fill valleys. Rocks, subjected to many kinds of erosion by water, are reduced to small particles and washed downhill. Rivers carry about 20 billion tons of eroded rock to the oceans every year.

Lakes form where water collects in depressions. Some lakes are very shallow, but the deepest are more than a mile deep. The majority of lakes contain fresh water, but the most brackish lakes are six times as salty as the ocean. Most lakes were created by glacial action. Notable among these are the Great Lakes of North America, where water has settled in depressions that were gouged out 20,000 years ago by glacial ice. Other lakes were once part of the ocean. Among these is the Caspian Sea, technically the largest lake in the world.

Rivers

The Earth's rivers contain more than 290 cubic miles (1,200 cubic km) of water flowing across the face of the Earth. Because they wear away at rocks and carry the debris to the ocean, rivers are among the Earth's most important agents of change.

Satellite images can be used to follow rivers from their snowy mountaintop sources, through broad fertile plains, past some of the world's great cities and on to the ocean. Most rivers begin their lives high in mountain areas, often as a small trickle flowing through inaccessible areas. As mountains give way to hills, rivers follow the topography, twisting and turning downhill on their way to the oceans. Clay and silt particles, collectively known as the sediment load, etch away at the sides and bottom of the river's channel. If a river contains more sediment than it can carry, the excess is dumped on either side of the river channel, creating meanders and building wide flood plains. The paths of past meanders create scars across the landscape. These can be identified from orbit by observing vegetation cover.

Usually, many small river tributaries join larger rivers on the way to the ocean. From space it is possible to map the layout of these drainage systems. The majority of river systems form dendritic patterns, with rivers laid out in shapes resembling the branches of a tree. Groundwater also enters the river, generally contributing between 10 and 30 percent of a river's flow.

Finally, rivers enter the oceans, usually forming deltas, as the sediment load spreads out over a wide area jutting into the sea. The shape of some deltas, such as the Mississippi, is determined by river flow. Other deltas, such as the Nile, are shaped by the movement of waves from the sea. Tidal forces dominate the deposition of many deltas, including the Mekong in Vietnam. Whatever form they take, all river deltas change over time.

Humans have used rivers throughout history. Most of the Earth's largest rivers have been dammed to control flooding, generate electricity and provide routes of transportation. These dams change the nature of rivers by trapping sediment loads and decreasing seasonal changes in flow. As rivers continue their way to the sea, they pass the great centers of human civilization. The first cities were built at the edge of rivers. Today, most of the world's great cities are adjacent to routes of water transportation.

Rivers often carve huge canyons. Some areas that contain the deepest canyons, such as the Grand Canyon in Arizona, are undergoing geological uplift — the land is being pushed upward. As this process continues, the river cuts its way through geological layers as they rise.

Some rivers carry large amounts of dirt and silt downstream. Faster-moving rivers carry much more sediment than slower-moving ones. If it were possible to double the flow rate of a river, it could carry 10 times as much sediment. Fast-moving flash floods have the power to carry boulders weighing tons. Rivers with

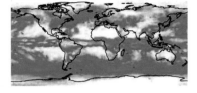
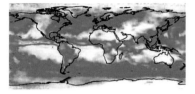

Measuring Rainfall

Precipitation can be measured accurately by rain gauges on the ground, but satellites are used to measure rainfall on the global scale.

Radar data from the Tropical Rainfall Measuring Mission satellite were used to produce these global maps of precipitation. The satellite uses radar to detect falling rain in the atmosphere. The maps show zones of low and high precipitation, including the Intertropical Convergence Zone near the equator. Tropical rainfall moves north to south each year in response to solar heating. These two maps were made six months apart to show the change in precipitation.

a high sediment load have an appearance very different from rivers with clear waters. A large amount of silt causes the river to appear much brighter than it would otherwise.

Many types of satellite images are well suited for observing and understanding rivers and the diverse areas through which they pass. Infrared images are used for detecting the presence of water, while wavelengths of blue light are effective for detecting the density of sediment in river waters.

The Amazon

The Amazon is by far the world's largest river in terms of volume. It carries about 20 percent of the Earth's flowing fresh water, more than that carried by the next seven largest rivers combined. The Amazon begins in the Andes Mountains in southern Peru, then crosses Brazil and empties into the Atlantic. The Amazon drainage system covers 2.3 million square miles (about 6 million sq km) and includes most of Brazil and Peru and parts of Colombia, Ecuador, Bolivia and Venezuela. The Amazon is about 4,000 miles (6,500 km) long. Only the Nile is longer.

For centuries explorers and geographers tried to find the most distant point in the Amazon drainage system, the source of the river. In the early 1700s three different parts of the Spanish colonial empire claimed that the great river began in their territory. By the mid-1900s it was generally accepted that the source was located in southern Peru. The use of air photos and satellite images in the 1970s helped explorers narrow down the search to a few small streams that form the headwaters of the Lloqueta River. In 2000 an expedition used Global Positioning System (GPS) equipment to map these streams accurately and locate the snow-capped mountain that is the farthest source of the Amazon drainage system.

Polar Ice

Seen from space, the Earth appears to be a bluish ball with white areas at the poles. These white areas are ice caps up to 10,000 feet (3,000 m) thick. Through millions of years the ice caps have expanded and retreated as temperatures rise and fall. Sheets of ice cover about 10 percent of the Earth's land surface and contain about 70 percent of the Earth's fresh water.

Recently, in geological terms, there was a period when the Earth had more ice than today. Until about 20,000 years ago ice covered North America as far south as New York City. The global climate is warmer today, so the great sheets of ice have retreated to polar areas. They did, however, have a profound effect on today's landscape. Before the last ice age the majority of rivers in North America flowed north to Hudson Bay. Today, most flow south into the Mississippi and on to the Gulf of Mexico.

Sheets of ice have an important effect on global climate. Ice reflects almost all the sunlight that reaches it, a phenomenon that has a cooling effect on the climate. More than 90 percent of the Earth's ice is in Antarctica. That continent is almost completely covered by a sheet of ice about 6,500 feet (2 km) thick in most places that continues to thicken by about 32 inches (82 cm) a year. The bedrock under the ice is barely above sea level. At the center of Antarctica, the ice is about 140,000 years old.

Geographers in Europe and the Middle East hypothesized about the existence of a southern continent, and Maori legends told of a long-ago visit to the frozen land to the south. However, no proof of such a land was found until 1820, when Antarctica was sighted from the sea. In the early 1900s, the heroic age of Antarctic exploration, several expeditions explored the Antarctic coast and the interior.

From 1908 until the outbreak of World War II, seven nations made claims to pie-shaped sectors of Antarctica, with the sectors meeting at the South Pole. Argentina, Chile and the United Kingdom disputed ownership of the Antarctic Peninsula; Australia claimed much of East Antarctica; New Zealand, Norway and France also made claims. Although competing claims led to friction, nations cooperated in scientific research during the International Geophysical Year in 1958, leading to the creation of several new research stations. An international agreement signed afterward set Antarctica aside for scientific uses, and the territorial claims have never been widely recognized.

The Antarctic ice sheet does not simply sit still; it slowly moves. At the South Pole stands a small monument near a research station. Every few years the monument needs to be relocated because the ice sheet under it moves by about 3 feet (1 m) annually. Near the coastlines, the ice moves faster as it flows off the mountains and downhill toward the ocean. Once it reaches the ocean, it floats as it continues to move outward. Eventually, pieces break off from the glacier to form icebergs. Many icebergs are small, but some are the size of a large city.

The continent itself moves over much longer time periods. Ice covers more than 99 percent of Antarctica, but millions of years ago it was closer to the equator, with a more tropical climate supporting forests and a diverse range of animal life. Since then continental drift has slowly pushed Antarctica toward the South Pole.

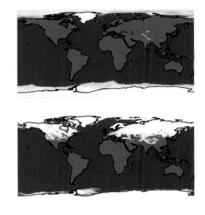

Snow and Ice

These global maps show the extent of snow and ice on the Earth. Snow and ice, because they are bright white, reflect almost all the sunlight that reaches them. This has important effects on global climate. Every year scientists track the amount of snow cover that changes in response to solar heating. These two maps were made six months apart. White indicates snow and ice cover, while land surfaces are shown as green.

There are places in Antarctica where lakes of liquid water exist under the ice sheet. The largest, called Lake Vostok, lies under a research station of the same name. This lake has been of interest to scientists studying the possibility of liquid water existing on Europa, an icy moon of Jupiter.

At the other end of the Earth at the North Pole, the Arctic Ocean is also covered by a shifting sheet of ice. Unlike at the South Pole, there is no land surface beneath the floating ice. As the ice moves around on the sea, it often breaks apart and refreezes. There is no monument at the North Pole, because the floating ice moves so quickly that no marker would stay in its place.

Greenland, the world's largest island, sits at the edge of the Arctic Ocean. Almost the entire island is covered by a layer of ice that averages about 5,900 feet (1,800 m) thick and contains about a million square miles (2.6 million sq km) of ice. This ice behaves much like the layer covering Antarctica, but on a smaller scale. This ice is so heavy that it actually weighs down the island, pushing most of the rocky surface of Greenland below sea level.

Scientists use radar and laser data from satellites to watch for changes in the thickness and movement of the ice sheet. They can detect changes in the height of the ice as small as a few inches. Measuring the thickness of the polar ice caps is important for determining how much water is stored in the Antarctic and Artic ice sheets, which hold most of the fresh water on the planet. As the ice melts water flows into the oceans, raising the sea level. Even a small change in the thickness of the ice sheet would have significant effects on the amount of water in the oceans. If all the ice in Greenland were to melt, sea level would rise by 20 feet (6 m).

Glaciers

Glaciers form where snow falls and does not melt. Newly fallen snow is not heavy; snowflakes contain about 90 percent air. However, as many years of snowfall accumulate the snow compacts into layers of solid ice containing less than 20 percent air. After the ice becomes about 100 feet (30 m) thick, it can no longer support its own weight and it slowly begins to flow downhill. Glaciers can flow about a foot every day, but some move much faster. The Jakobshavn Glacier in Greenland moves almost a meter every hour.

By trapping air as they form, glaciers can record important information about the climate. Scientists take samples of ice laid down thousands of years ago to determine the temperature, gases and amount of oxygen that once existed in the atmosphere. This information has played an important part in helping us understand how the Earth's climate changes through time.

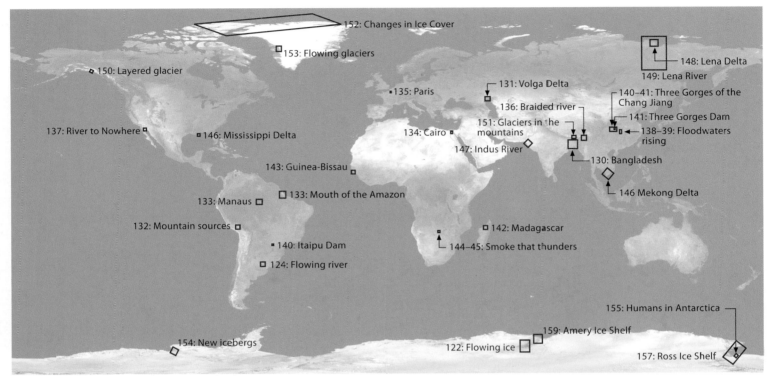

152: Changes in Ice Cover
153: Flowing glaciers
150: Layered glacier
135: Paris
131: Volga Delta
136: Braided river
151: Glaciers in the mountains
148: Lena Delta
149: Lena River
140–41: Three Gorges of the Chang Jiang
141: Three Gorges Dam
138–39: Floodwaters rising
137: River to Nowhere
146: Mississippi Delta
134: Cairo
147: Indus River
130: Bangladesh
143: Guinea-Bissau
133: Mouth of the Amazon
146 Mekong Delta
133: Manaus
132: Mountain sources
142: Madagascar
144–45: Smoke that thunders
140: Itaipu Dam
124: Flowing river
155: Humans in Antarctica
154: New icebergs
159: Amery Ice Shelf
122: Flowing ice
157: Ross Ice Shelf

Numbers indicate page locations of the images in this chapter.

As they flow downhill, glaciers push dirt, rocks and everything else out of the way, eroding the land and leaving distinctive marks on the landscape. Glaciers create wide U-shaped valleys, distinct from the V-shaped valleys created by rivers. Some of the most impressive canyons on Earth, such as the Yosemite Valley, have been formed by glaciers. Glaciers carry large rock fragments, which carve long, straight grooves as they pass over the land. Huge piles of dirt, called moraines, accumulate at the sides and margins of glaciers.

Because glaciers are made of ice they have another characteristic: they melt. While new snow falls on mountaintops to form more ice, a glacier melts when it reaches lower altitudes. Whether the glacier is getting larger or smaller depends on which process is more rapid. During the past century the majority of glaciers, including all the large glaciers in coastal Alaska, shrank. Scientists are studying how glacial melting is related to rising global temperatures, and satellite images are being used to detect changes in the size of glaciers.

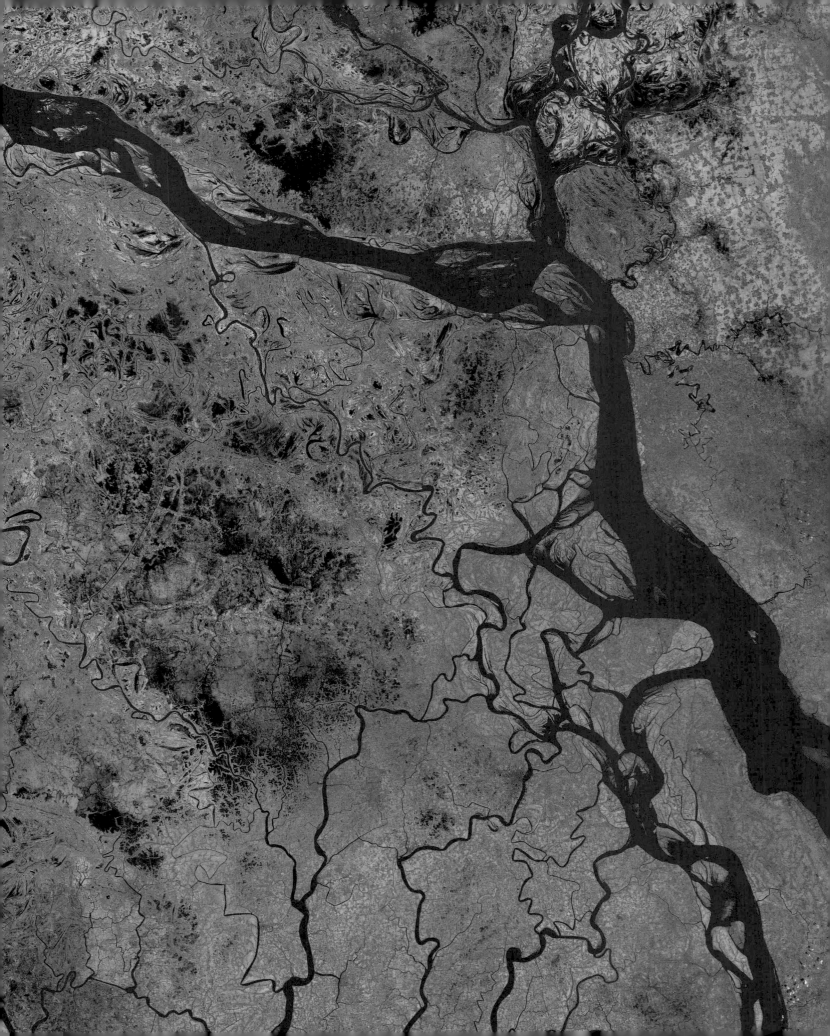

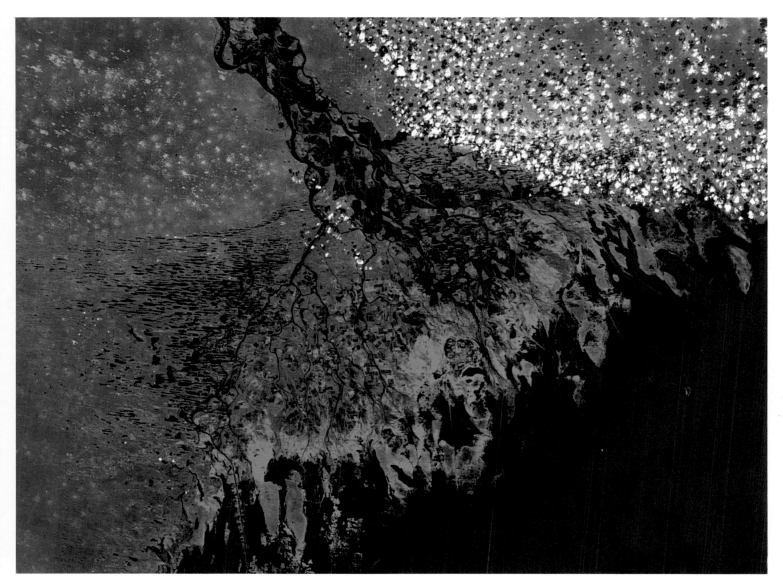

Living with Changing Rivers

BANGLADESH (LEFT)

Perhaps no other country is as dominated by rivers as Bangladesh. Waters from the Ganges and Brahmaputra rivers, as well as hundreds of other channels, flow through Bangladesh into the Indian Ocean. Most of Bangladesh is less than 340 feet (110 m) above sea level, allowing the country's great rivers to flood and change their course frequently. In 1787 a large flood altered the Brahmaputra to the course it takes today. Every year during the rainy season the rivers rise, making Bangladesh extremely prone to flooding damage.

This image from the Landsat 7 satellite covers about one-quarter of Bangladesh. The Ganges and Jamuna rivers join just off the upper-left corner, forming the Padma River. Dhaka, the capital and largest city, lies in the northern part of the image. Along the coast, the river channels allow for the growth of dense wetland forests, known as the Sundarbans, shown here as the bright green area to the south.

VOLGA DELTA (ABOVE)

The Volga, the longest river in Europe, flows about 2,200 miles (3,700 km) from hills near Moscow to the Caspian Sea. Its source is less than 800 feet (240 m) above sea level, so the river flows gently to the sea. Temperatures in the area are generally cold and the river is frozen for more than five months each year. This image, acquired by the Resurs-O1 3 satellite on June 18, 1997, shows the delta of the Volga River, where it empties into the Caspian Sea. The bright green color indicates dense vegetation.

Channels dug through the Volga Delta to allow ships to pass are visible as straight lines running to open water. The damming and channeling of the Volga has created economic growth for Russia, but it has also decreased the river's flow and adversely affected fish. Before the dams were built, Caspian Sea sturgeon, the source of Russian caviar, swam up the Volga to spawn. Today most sturgeon are raised artificially.

Dhaka →

N

Ganges (Padma) River

Brahmaputra River

Meghna River

50 Km

Sundarbans

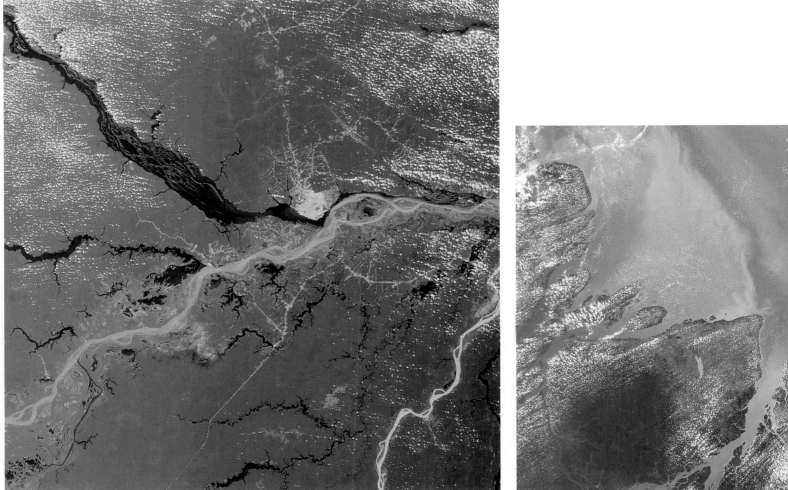

Amazon from Beginning to End

MOUNTAIN SOURCES, PERU (LEFT)

This photograph, taken by Space Shuttle astronauts in October 1988, shows the farthest source of the Amazon drainage system — an area north of the city of Arequipa in southern Peru. A few large volcanoes are visible. To the south is Ampato, which rises over 20,000 feet (6,310 m) above sea level. Next is Sabancaya, with dark lava flow on its east side. Near the center of the image is the Colca Canyon, which is much deeper than the Grand Canyon in Arizona. To the north of the Colca, on the other side of the Cordillera de Chilca, five small streams join to form the Lloqueta River. Flowing north, this stream is the farthest source of the Amazon.

MANAUS, BRAZIL (ABOVE LEFT)

The Amazon, a broad and slow-moving river, is more than 100 feet (30 m) deep throughout Brazil and flows at only about one-and-a-half miles (2.5 km) an hour. The elevation of the river is about 300 feet (90 m) above sea level at the Brazil-Peru border. Oceangoing ships can sail up the Amazon as far as Iquitos, Peru.

This image was collected by the MISR sensor on the Terra satellite on July 23, 2000. It shows the point where the Negro and Solimões rivers join to form the Amazon at Manaus, the largest city within the Amazon basin. The Negro, the largest river to join the Amazon, accounts for about 20 percent of the Amazon's volume. Other large tributaries include the Madeira in Brazil and the Marañón and Ucayali in Peru. Rivers that have their source in the Andes are laden with mountain sediments, giving the water a brighter color in this image. The Negro carries little sediment, so its clearer water appears darker. The water from the two rivers does not completely mix until far downstream.

MOUTH OF THE AMAZON (ABOVE RIGHT)

At the point where the Amazon reaches the Atlantic, a vast amount of fresh water and sediment mix with ocean water. The mouth of the Amazon is hundreds of miles across. About 6.2 million cubic feet (175,000 cubic m) of water flow through the mouth of the Amazon every second, roughly 10 times the flow for the Mississippi River. The Amazon empties about 1.3 million tons of light-colored sediment into the ocean every day. As the fresh river water mixes with the ocean, the saltiness of the ocean is diluted more than 100 miles (160 km) offshore. This image, collected by the MISR sensor on the Terra satellite on September 8, 2000, covers an area about 235 miles (380 km) wide.

Great Cities on Rivers

PARIS (RIGHT)

Many of the world's greatest cities have grown up around rivers, which serve as transportation links fueling the growth of commercial centers. The Seine is France's second-longest river, after the Loire. Paris, the most important city on the Seine, is shown in this high-resolution image. At the center of this image, acquired by the QuickBird satellite on March 27, 2002, is the Eiffel Tower, one of the most famous landmarks in the world. Designed by Gustave Eiffel and made of wrought iron, the tower was built in 1889 for the International Exhibition. At 300 meters (984 feet), it surpassed Washington DC's Washington Monument as the tallest structure in the world and held that status until New York's Chrysler Building eclipsed it in 1930.

CAIRO (LEFT)

The Nile, about 4,100 miles (6,800 km) long, is the longest river in the world. Far to the south the Nile is formed by the union of the White and Blue Niles at Khartoum, Sudan. Water flowing down from the highlands of central Ethiopia forms the White Nile, while the sources of the Blue Nile lie in the mountains around Lake Victoria in the African Rift Valley.

This image, collected by the ASTER sensor on the Terra satellite on April 1, 2003, shows Cairo, Egypt's capital and largest city. The city lies at the point where the Nile River begins to branch out into a wide delta on its way to the Mediterranean Sea. In the 1800s a new Cairo was built north of the old city, influenced by city-planning ideas used in Paris. Modern Cairo, spreading out across both sides of the Nile, is a fast-growing city, with new developments pushing farther into the desert. Vegetation of the Nile River Valley appears red in this infrared image.

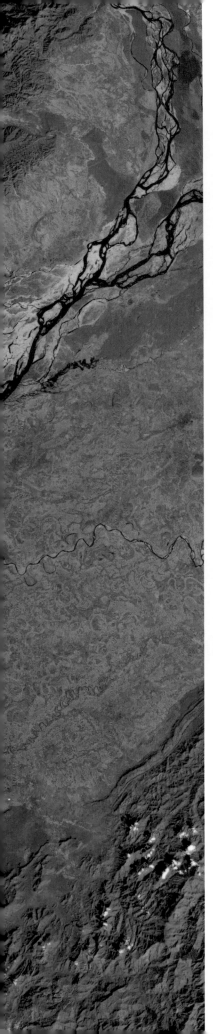

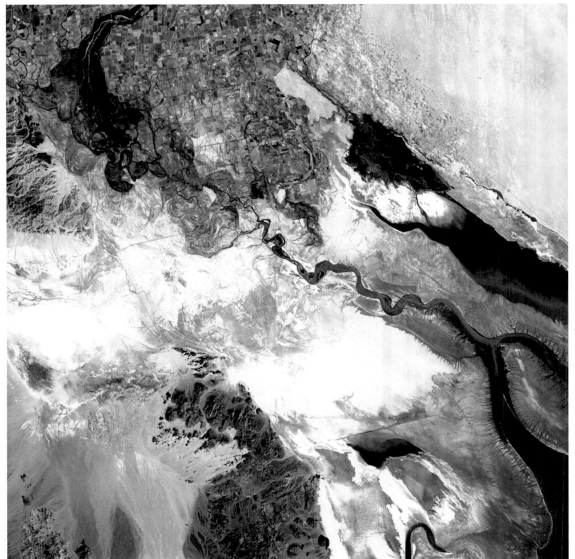

Braided River

BRAHMAPUTRA RIVER (LEFT)

This image, acquired by the Landsat 5 satellite on January 14, 1989, shows the Brahmaputra River as it winds through the northeastern corner of India. The river has many braided channels that change shape as water levels rise and fall. Green indicates vegetation in the mountains to the northwest. The braided river channels are shown in blue, and the floodplain appears in a lighter color.

The Brahmaputra begins its journey in the glaciers and mountains of southwestern Tibet, where it is widely known as the Tsangpo River. As the Tsangpo passes through the Assam Valley into India, it is joined by the Dibang and Lohit rivers to form the Brahmaputra. Part of the confluence of the three rivers is visible in the upper right of this image. The Brahmaputra eventually joins the Ganges in Bangladesh and empties into the Bay of Bengal.

River to Nowhere

COLORADO RIVER DELTA (ABOVE)

The image above was collected by the ASTER sensor on the Terra satellite on September 8, 2000. It shows the Colorado River as it approaches the Gulf of California in northern Mexico. In the past the river emptied into the gulf. Beginning in the 1930s significant amounts of water from the Colorado River were diverted for irrigation and drinking water. This made it possible to develop large sections of the southwestern United States, but today only about 10 percent of the river's original flow reaches Mexico. Except during flood events, the river dries up near the center of this image. The blue and purple colors in the lower right of this infrared image show the former river channel, which is now occupied by brackish water from the Gulf.

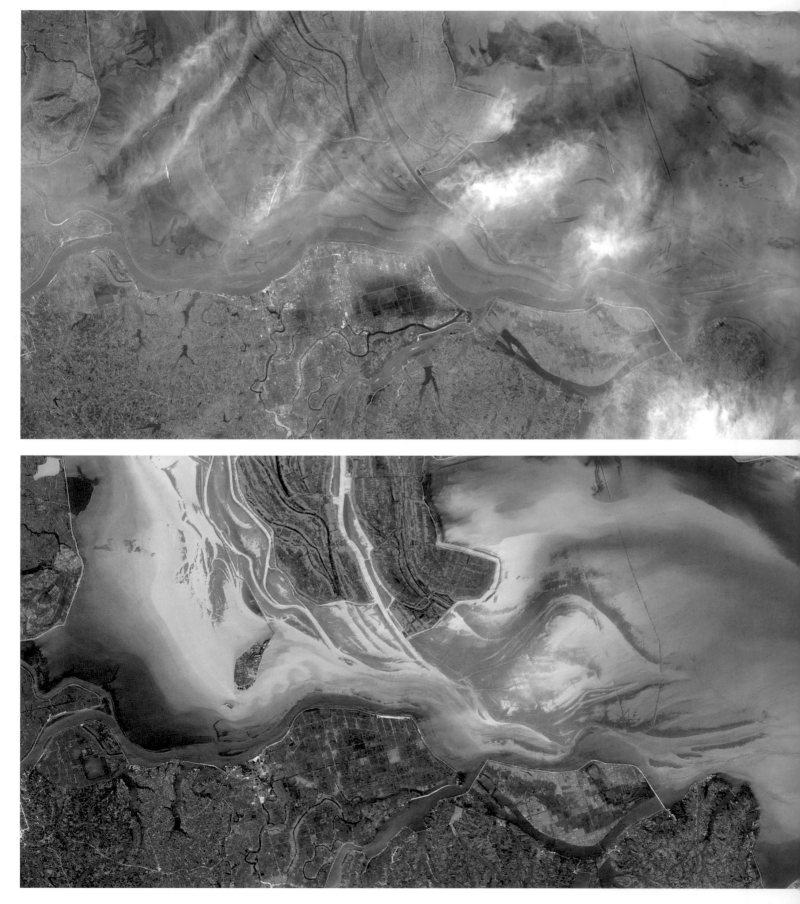

Floodwaters Rising

DONGTING LAKE, CHINA
MARCH 2002 (TOP LEFT)
SEPTEMBER 2002 (BOTTOM LEFT)

Rivers do not always stay within their boundaries. Floods can occur when precipitation is high, causing widespread changes in the local environment and devastation to humans. These images show Dongting Lake in northern Hunan Province of China. The lake sits on land created by sediments washed downstream by the Chang Jiang, also called the Yangtze River. Every summer the level of the Chang Jiang rises and a network of pumps, dikes and channels divert floodwaters into holding ponds.

These satellite images were taken by the ASTER sensor on the Terra satellite six months apart, in March and September 2002. The first image shows the lake before flooding, under hazy atmospheric conditions. In August, the Chang Jiang and other rivers flooded and carried enormous volumes of water downstream into the lake, covering low-lying areas.

Controlling Rivers

Itaipu Dam (above left)

Humans have built dams across most of the world's large rivers to control flooding, provide ship channels and produce electricity. This high-resolution image from the IKONOS satellite provides a view of the huge Itaipu Dam on the border between Paraguay and Brazil. An important source of electricity for Paraguay, water from the Paraná River spins 18 turbines (located next to the base of the dam near the top of the image). Spillways, visible on the left, allow excess water to be released when necessary.

Three Gorges of the Chang Jiang (Top Right)

The Chang Jiang, also known as the Yangtze River, is the longest river in Asia and has been an important route for Chinese civilization for thousands of years. Beginning high in Tibet, the Chang Jiang flows through narrow valleys before becoming more than 1,000 feet (300 m) wide at lower altitudes. The river then flows through the Three Gorges in central China. The gorges are steep canyons with walls rising nearly vertically from the river to a height of almost 2,000 feet (600 m). At about 600 feet (180 m), the Chang Jiang was the deepest river in the world before it was dammed in the 1990s.

This image was collected by the ASTER sensor on the Terra satellite in July 2000. It shows about 35 miles (60 km) of the Chang Jiang as it flows from left to right through the Three Gorges, toward the Pacific Ocean. The Three Gorges Dam is under construction on the left, and the earlier Gezhouba Dam is downstream. In the center of this image is the Xiling Gorge. The other two gorges, upstream from the dam, are being submerged as the reservoir fills.

Building the Three Gorges Dam (right)
Three Gorges Dam in Operation (far right)

The dam is seen under construction on April 21, 2002, in this image from the QuickBird satellite (right). Ship channels are under construction along the top. The first oceangoing ships passed through the channels in June 2003. The image on the far right, acquired by the QuickBird satellite on July 13, 2003, shows the dam beginning to hold back water.

Large dams have been built in China for thousands of years. A stone dam about 100 feet (30 m) high was built across the Gukow River in 240 BC. The Three Gorges Dam, at about 8,000 feet (2,400 m) long and 610 feet (185 m) high, controls flooding and generates electrical power for China's growing cities. However, it also radically changed the landscape. The reservoir created by the dam will be about 360 miles (600 km) long when filled by 2012. More than a million people were relocated to higher ground as the reservoir filled, including about 100,000 residents of the city of Fengjie, about 150 miles (240 km) upstream from the dam.

Three Gorges Dam

Chang Jiang

IMAGE ON LEFT

IMAGE BELOW

10 Km

Gezhouba Dam

N

Rivers Carry Sediment to the Sea

MADAGASCAR (ABOVE)

The Betsiboka River, one of several large rivers on the western side of the island of Madagascar, flows from the mountains to the Indian Ocean. At the mouth of the Betsiboka is the city of Mahajanga. In the 1700s this was the capital of the kingdom of Boina, covering most of northwestern Madagascar. In 1895 the French military landed at Mahajanga, and the French controlled the island until the Malagasy Republic gained independence in 1958. Today's culture combines influences from Asia and Africa.

As the mountains of Madagascar are slowly pushed upward by geological forces, rivers such as the Betsiboka erode them away. Large amounts of sediment carried by the river cause the distinctive braided channels visible at the river mouth in the northwest corner of the image. Intensive use of the land has allowed a high amount of erosion, adding to the sediment load in the Betsiboka. This image is a combination of data collected by the Landsat 5 and SPOT 4 satellites in 1986–87.

GUINEA-BISSAU (RIGHT)

The Geba River flows through Guinea-Bissau in western Africa on its way to the Atlantic Ocean. Rivers meander across plateaus in the interior of the country, then widen into broad inlets as they approach the ocean. Carrying loads of silt, the rivers create the swirls in the ocean visible here. Bright red indicates the infrared reflection of vegetation on the land. Bissau, the capital and largest city in the country, lies on the Geba River to the right. This image was collected by the Landsat 7 satellite on December 1, 2000.

Smoke That Thunders

VICTORIA FALLS (RIGHT)

Victoria Falls is the largest waterfall
in the world. Located on the border
between Zambia and Zimbabwe and
originally known as Mosi-oa-Tunya
("the smoke that thunders"), the falls
are over 300 feet (90 m) high and about
5,500 feet (1,700 m) wide, and generate
a plume of water visible for more than
12 miles (20 km). Fish unlucky enough
to go over the falls are often killed, and
otters wait at the base to take advantage
of an easy meal.

The Zambezi River flows from
Angola and northwestern Zambia to
the Indian Ocean. In the area around
Victoria Falls, the river eroded one of the
natural joints in the volcanic rock and
widened it into this major waterfall. In
this image, acquired during the rainy
season by the QuickBird satellite on
March 26, 2002, the Zambezi River exits
to the left. Flow through the river drops
by 90 percent during the dry season in
October. A tourist hotel is visible in
the lower right.

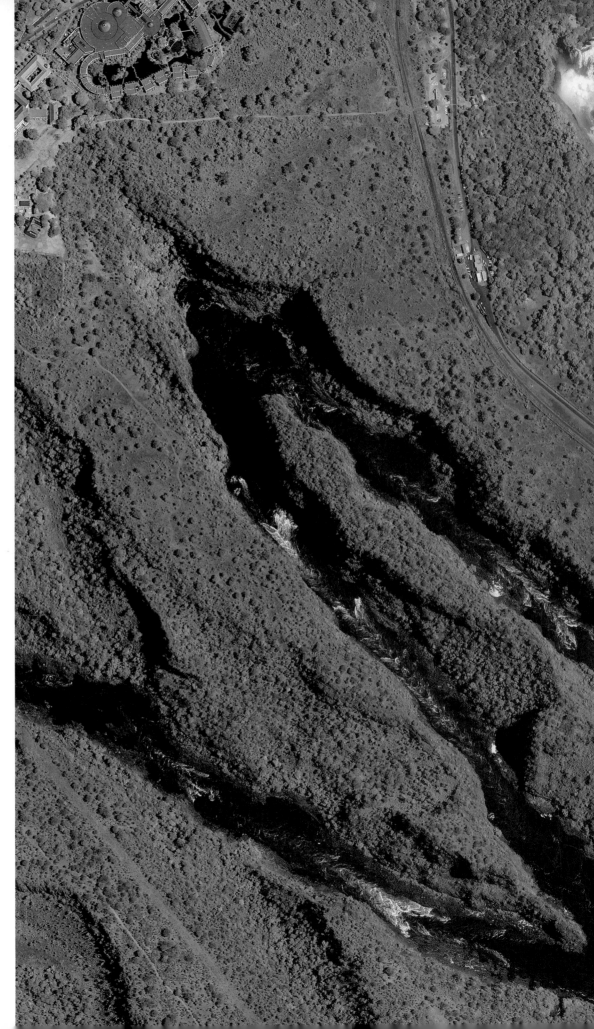

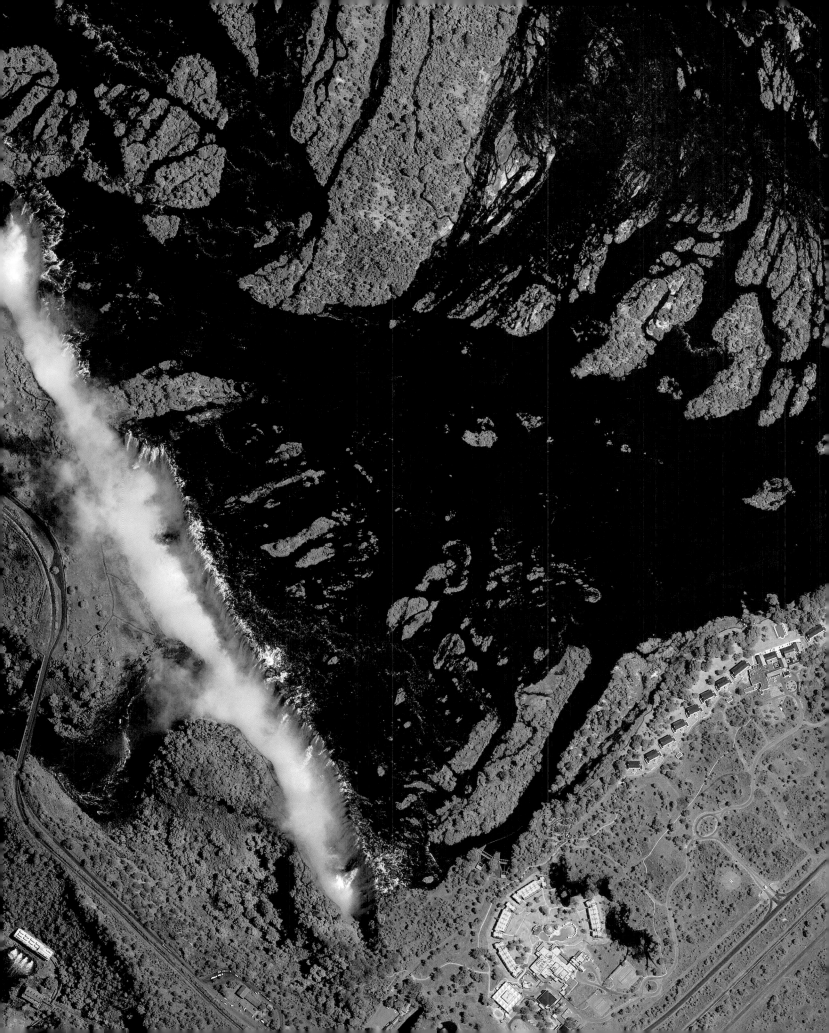

River Deltas

MEKONG (ABOVE LEFT)

The Mekong, the longest river in southeast Asia, starts its journey high in the mountains of central China. It flows about 2,600 miles (4,160 km) through China, along the borders of Laos, Burma and Thailand, and continues through Cambodia and southern Vietnam to empty into the South China Sea. The Mekong carries great quantities of mud and silt into the ocean. Currents carry this material to the south, visible in the photograph as a long plume in the ocean.

The Mekong Delta, a huge landmass about 100 miles (160 km) wide, is home to one of Asia's most productive areas of rice cultivation. Space Shuttle astronauts took this photograph in February 1996. The delta is visible in the center. Channels have been dug to irrigate rice fields in the dry winter months; these are visible through the reflected sunlight on the right side of the image. Ho Chi Minh City (formerly Saigon), the largest city in Vietnam, is at the right edge of the image.

MISSISSIPPI (ABOVE RIGHT)

The Mississippi River, the largest river in North America, flows throughout the central United States, traveling more than 3,000 miles (5,000 km) before it reaches the Gulf of Mexico. In the middle of its course, the river meanders over flatter, easily eroded terrain. About 600,000 cubic feet (17,000 cubic m) of water leave the Mississippi every second, carrying about 220 million tons of sediment downstream every year. During the last six thousand years, the river has deposited the sediment over a delta area 200 miles (320 km) long.

This image of the Mississippi River Delta was collected by the ASTER sensor on the Terra satellite on May 24, 2001. Green shows vegetation on the lands of the delta. The lighter colors in the water show silt from the river mixing with the Gulf of Mexico. Although the river is continually changing course through the delta, the central channel is kept open for shipping. In the upper part of the image, ship wakes appear on the water.

INDUS (RIGHT)

The Indus River begins in the Himalayas and flows through northern India, but most of its length is in Pakistan. The Indus Valley has long been an important center of human culture. The oldest settlements found by archaeologists date back 7,000 years. By 2600 BC, large cities existed near the Indus River in modern Pakistan and India.

This photograph shows the delta of the Indus in Pakistan, where the river divides into many small channels before it empties into the Arabian Sea. The main channel is visible in the foreground. In the background is the Rann of Kachchh, part of which can be seen on page 137. Marshy areas flood every year during the monsoon season beginning in June or July. The course of the Indus has been slowly shifting to the west. About 200 years ago it emptied into the sea through the Rann of Kachchh, but today's course takes it directly through the channels seen here. This photograph was taken by astronauts on the Space Shuttle on March 2, 1996.

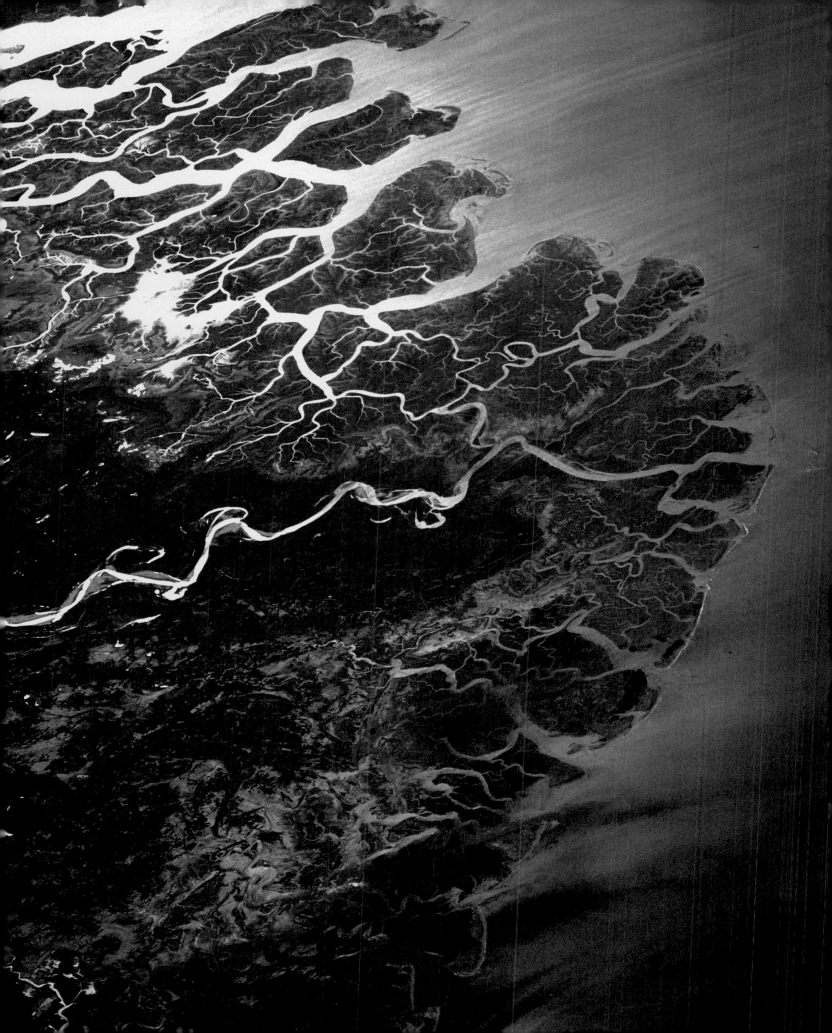

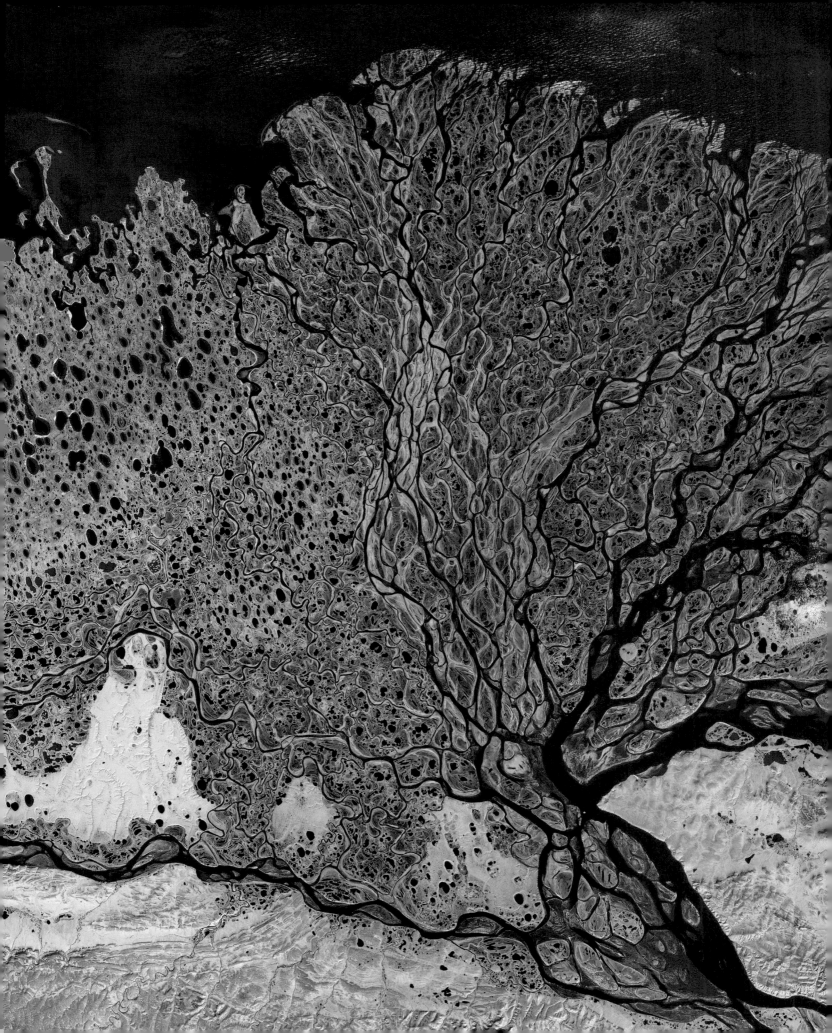

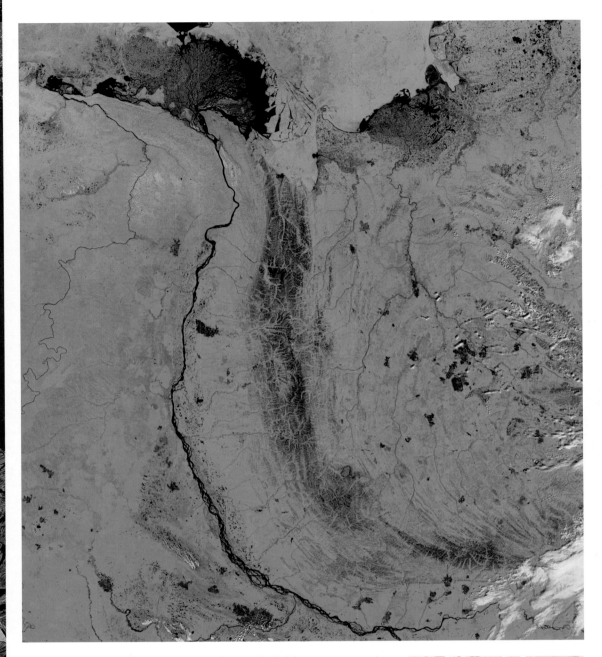

Flowing to the Arctic

LENA RIVER DELTA (LEFT)

LENA RIVER (ABOVE)

The Lena River flows north for over 2,800 miles (4,500 km) through Russian Siberia to the Arctic Ocean. As the Lena enters the ocean, it divides into many channels and forms a large delta. The Lena Delta is a protected reserve home to a wide range of wildlife. The Landsat 7 satellite acquired the image on the left in July 2000.

The lower half of the Lena is seen in the image above, acquired by the MODIS Sensor on the Terra satellite on June 28, 2002. No dams or channels have been built here, so the river meanders across a flat floodplain several miles wide. Fires occur every year in Siberia, and reddish burn scars from forest fires can be seen. Snow and ice are visible on the mountain peaks, which reach a height of 7,000 feet (2,100 m). Around the delta, sea ice breaks up during the annual spring thaw.

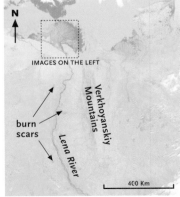

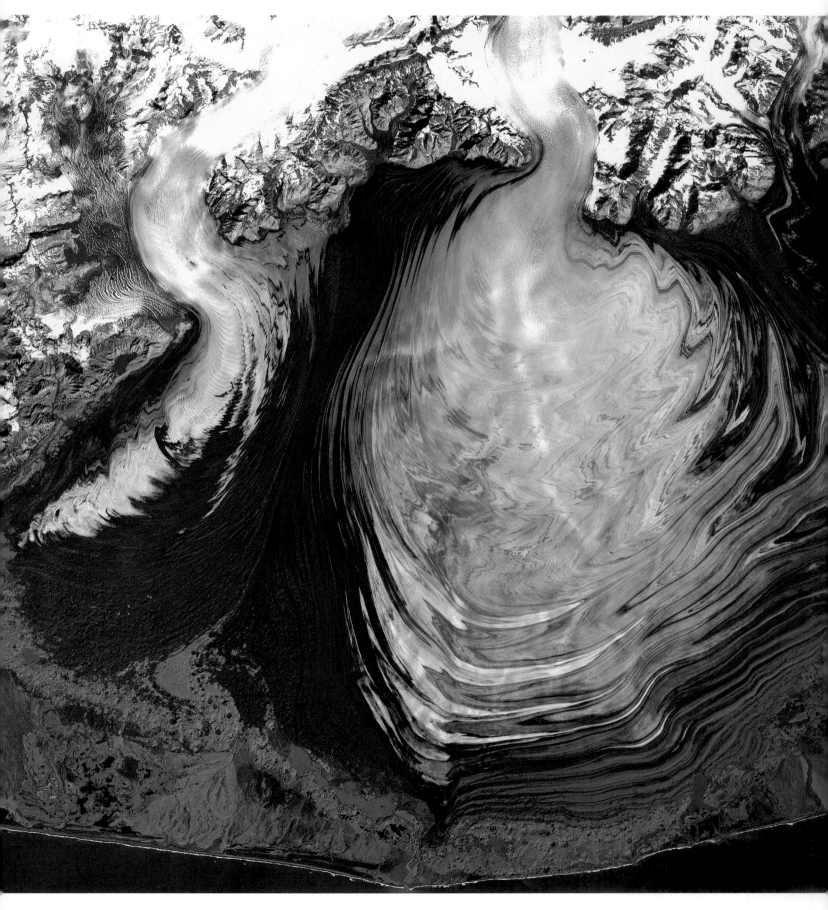

Layered Glacier

MALASPINA GLACIER, ALASKA (LEFT)
This image shows the Malaspina Glacier, a river of moving ice up to 2,000 feet (600 m) thick. Covering 1,500 square miles (3,900 sq km), it is the largest glacier in Alaska. Glaciers are made up of many layers of ice, which become exposed as the glacier melts. This image, acquired by the Landsat 7 satellite on August 31, 2000, shows the layers as alternating white and red stripes near the margins of the glacier. Vegetation thrives around the edges of the ice, which appears brown in this infrared image. The slopes of Mount Saint Elias, the second-highest peak in the United States, are visible in the northwest corner of the image. Only 11 miles (18 km) from the Pacific coast, the mountain rises more than 18,000 feet (5,500 m) above sea level. The area is part of the Wrangell–Saint Elias National Park and Preserve, the largest National Park unit in the United States.

Glaciers in the Mountains

HIMALAYAS, BHUTAN (ABOVE)
Most people in Bhutan live in agricultural communities in the valleys and plains to the south. Until recent times, Bhutan's mountainous location kept the county in relative isolation. Before 1960, traveling to the capital, Thimphu, required a six-day trip by foot or mule on dirt paths. Today, paved highways cross the western part of Bhutan.

The northern border of the country is defined by the Himalayan Mountains. This image, acquired by the ASTER sensor on the Terra satellite on November 20, 2000, shows the Himalayas covered by snow and ice, as glaciers flow down both sides of the mountains. At the end of many glaciers are small lakes created from melting ice. These glaciers, like many around the world, are melting and shrinking.

Greenland

CHANGES IN ICE COVER (LEFT)

Permanent ice sheets cover the Earth's north and south poles. About 85 percent of the island of Greenland is covered by sheets of ice as thick as 9,900 feet (3,000 m), the largest reservoir of fresh water outside Antarctica. The weight of the ice is so great that it presses much of the underlying rock below sea level.

This image, acquired by the MODIS sensor on the Terra satellite on July 3, 2002, shows the northern edge of Greenland and Ellesmere Island, Canada (north is to the left in this image). Changes in ice cover are measured using data from satellites. Melting of Greenland's ice cap may have a significant impact on sea level changes. New islands have appeared along some of the coastlines of Greenland as ice cover shrinks.

FLOWING GLACIERS (RIGHT)

Shown here is the western coast of Greenland, where large glaciers are flowing down off the land and onto the waters of Baffin Bay. The dark brown areas are rocky outcrops poking through the ice. This image was collected by the Landsat 7 satellite on September 3, 2000.

It is on the western coast of Greenland that human settlements have taken root. Norse settlers from Iceland arrived on the western coast of Greenland in 982 AD. The last written record from these settlements — cooling temperatures may have been an important factor in their decline — dates from 1408. Today Greenland is Danish territory with a level of autonomy.

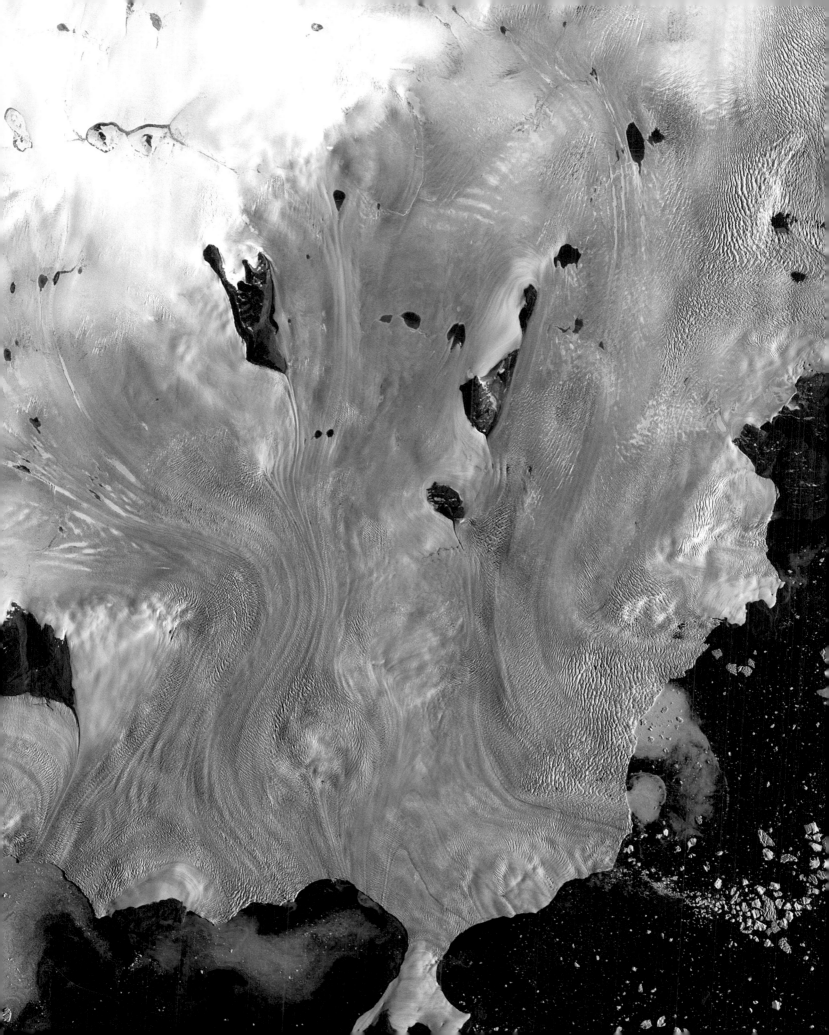

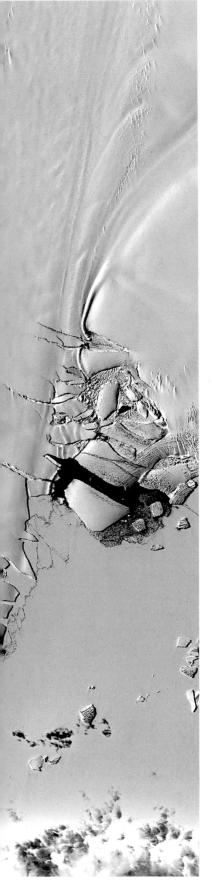

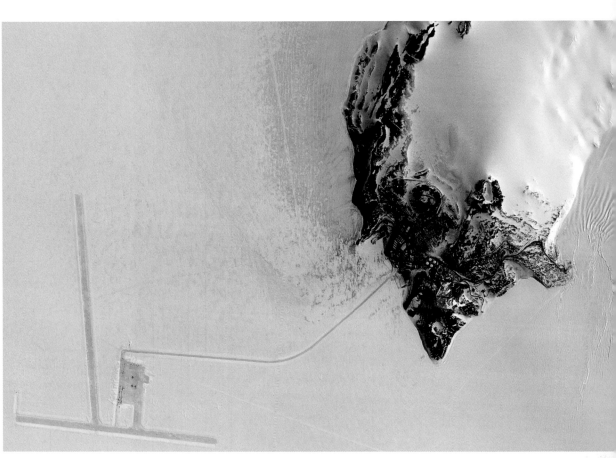

New Icebergs

PINE ISLAND GLACIER, ANTARCTICA (LEFT)

The Pine Island Glacier in Antarctica is shown here, floating on the surface of the ocean. The point where the ice loses contact with the land is called the grounding line. As the ice moves far beyond the line, it starts to break off from the glacier.

In this image, acquired by the ASTER sensor on the Terra satellite on December 12, 2000, a large piece of ice is in the process of calving, or breaking away, from the glacier. The wide crack running across the image was identified — and the calving predicted — through satellite imagery. The ice finally broke free 11 months later, and a new iceberg floated out to sea.

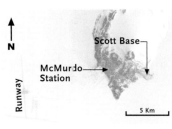

Humans in Antarctica

MCMURDO STATION (ABOVE)

Antarctica does not have a permanent human population. In the 1950s many nations claimed parts of the continent, but those claims were not enforced or widely recognized. Many tourists and thousands of scientists visit Antarctica each year.

This image from the QuickBird satellite shows McMurdo Station, the largest settlement on the continent, with a summer population of about 1,000. Most people leave in the winter. McMurdo, the headquarters of the

U.S. scientific operations, was built in 1955 on Ross Island. Rocky terrain not covered by ice appears brown in this image. Out on the smooth ice sheet, airstrips have been plowed and three large aircraft are visible. Buildings and fuel tanks are also visible in the station itself.

McMurdo Station lies at the world's southernmost point of exposed land. Robert Scott began his journey from this area in an unsuccessful attempt to beat Roald Amundsen to the South Pole. His camp remains, with supplies from 1910 still sitting on the shelves. New Zealand's Scott Base is nearby.

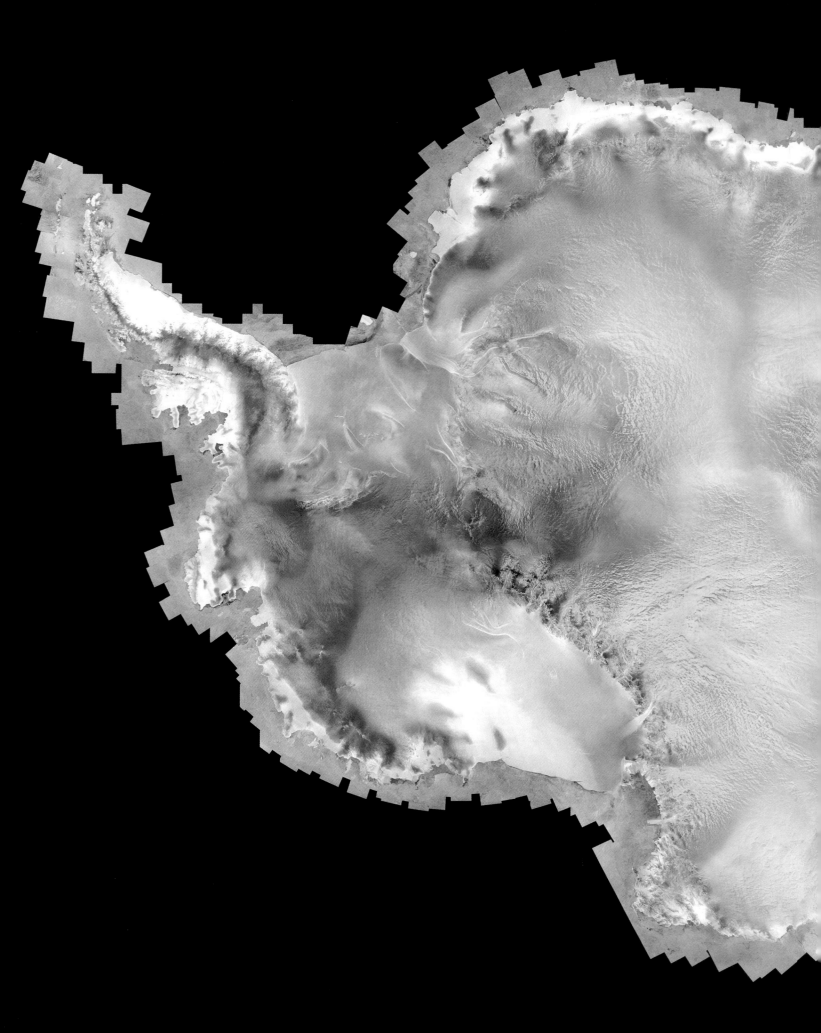

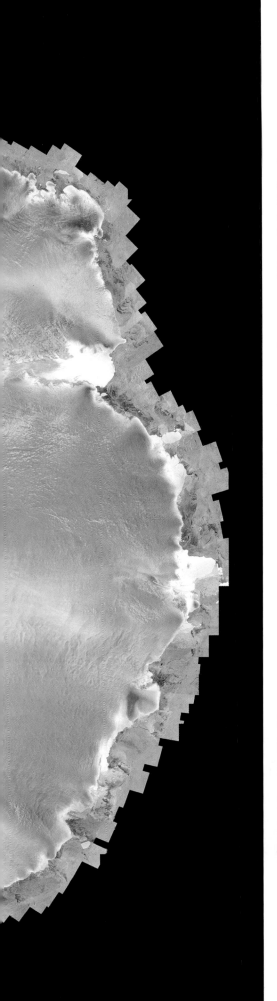

The Frozen Continent

ANTARCTICA (LEFT)
ROSS ICE SHELF (RIGHT)

Antarctica lies at the southern end of the world, with the South Pole roughly in its center. The continent is divided into two parts by the Transantarctic Mountains. West Antarctica includes land around the Ross and Weddell seas. The Antarctic Peninsula reaches north to within 600 miles (1,000 km) of South America. East Antarctica is covered by the thickest ice sheet in the world

A radar image of the entire continent (left) is able to reveal the texture of the ice surface, and it shows detail that is not possible to obtain with visible or infrared sensors. It was produced from data returned by the RADARSAT-1 satellite. The image on the right, created from the same data, shows the western margin of the Ross Ice Shelf. Individual glaciers are visible moving out to sea.

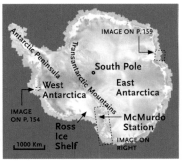

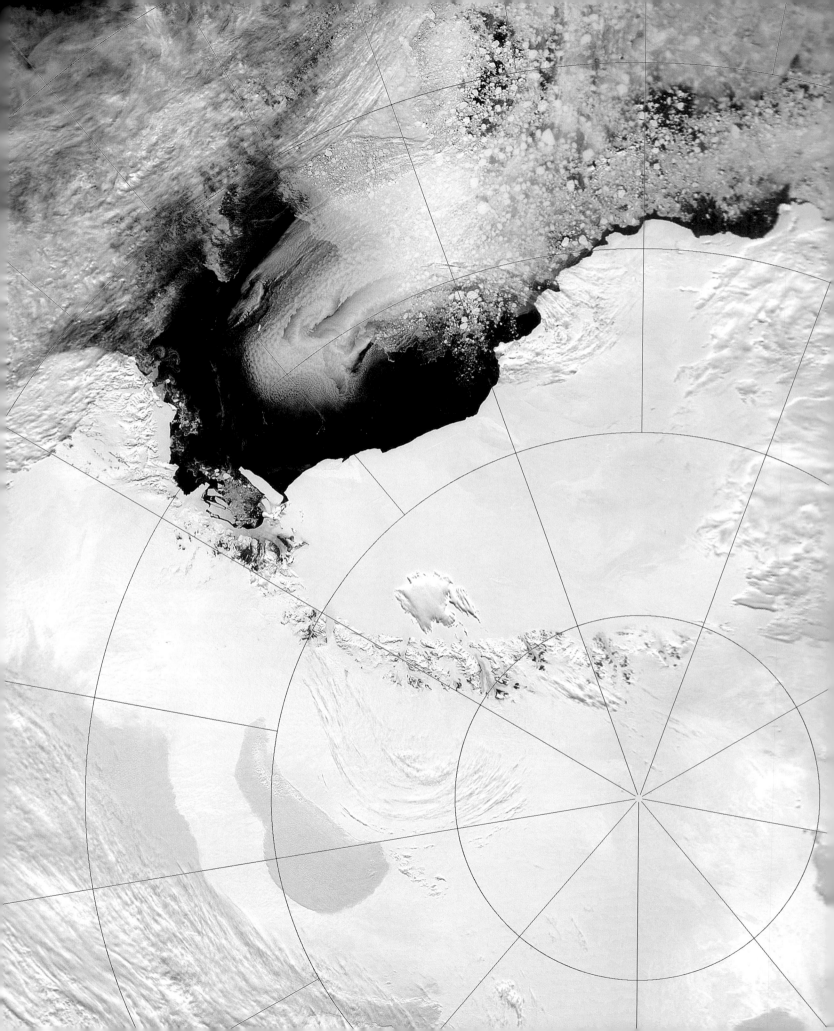

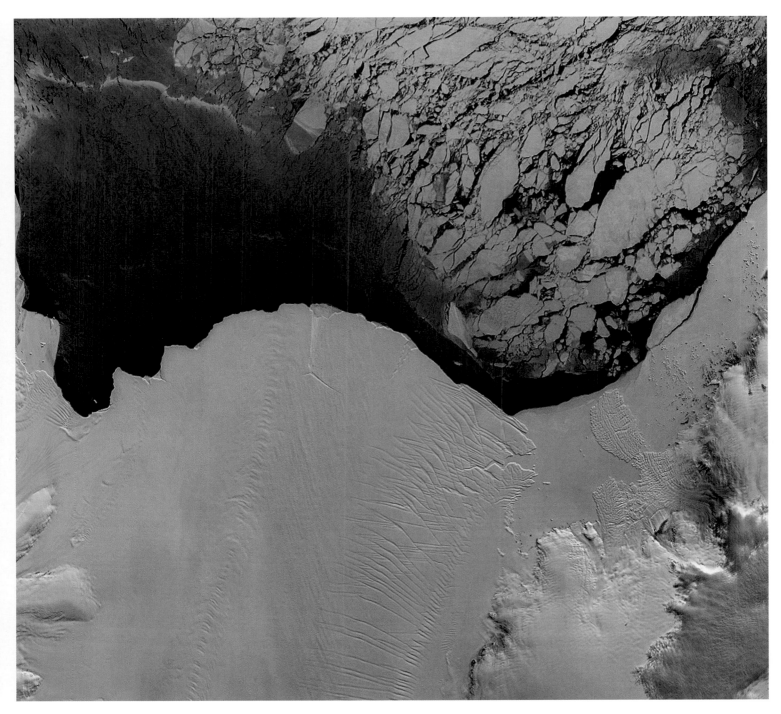

The Frozen Continent

THE SOUTH POLE (LEFT)

This image shows a rare clear view of the South Pole and the Ross Ice Shelf. The Ross Ice Shelf is a massive sheet of ice floating on the ocean. It lies about 1,000–2,000 miles (1,600–3,200 km) north of the South Pole. The ice floating in McMurdo Sound has recently broken off as new icebergs that are floating out to sea. The water off the coast appears green, indicating a high density of phytoplankton.

Latitude and longitude have been plotted on the image, acquired by the SeaWiFS sensor on the OrbView-2 satellite. Lines of longitude converge at the South Pole. The South Pole is at 9,300 feet (2,900 m) above sea level.

AMERY ICE SHELF (ABOVE)

This October 6, 2001, image from the MISR sensor on the Terra satellite shows the Amery Ice Shelf in East Antarctica. Large fractures up to 12 miles (20 km) long can be seen in the ice. When these cracks extend across the ice a new iceberg will be formed. The MISR sensor has nine separate cameras that view the Earth at different angles. By combining data from these cameras it is possible to discriminate clouds from ice and detect small features on the ice. Smooth surfaces are blue, rough features are red or orange, and clouds are purple.

The Human Presence

T he surface of the Earth shows the impact of humans building, moving and changing things to create new opportunities for themselves on this planet. Although less than 3 percent of the Earth's land surface is occupied by features built by humans, these areas are spread over the entire planet. From our footprint on the ground to the lights that brighten our cities, the human presence all over the world is plain to see from space.

Satellites can also observe some of the unintended consequences of the presence of humans. Global-scale consequences of pollution, land use and water use are the focus of several remote sensing research projects.

Looking into the Past

Most structures built by the earliest societies were small and, if still standing, would not be visible from space. There are a few exceptions, such as the pyramids at Giza, Egypt, and the Great Wall of China. It is often stated that the Great Wall of China is the only human-made thing visible from space, but that is not true. An astronaut in orbit would probably not be able to see the Great Wall with unaided eyes. It is also sometimes said that the Great Wall is visible from the Moon — another statement that is not true. To an observer standing on the Moon, the Earth is larger than the Moon appears in our sky, but still easily obscured by a person's thumb at arm's length.

Archaeologists often use satellite images to locate areas where ancient buildings and trade routes existed. In forested regions, stone foundations leave a mark on surface vegetation. Ancient trade routes have been mapped from space by identifying paths where the soil has been compacted by legions of travelers. With satellite images it is possible to scan large areas, allowing sites to be located more efficiently.

Ancient Monuments
GIZA, EGYPT

Still among the largest structures ever built, the pyramids at Giza, Egypt (opposite page), are over 4,000 years old. They were built as tombs and monuments for the rulers of Egypt. The Khufu Pyramid, the largest and oldest, lies to the north in this image. Originally more than 480 feet (145 m) high and made of 5.7 million tons of limestone, it was named for the ruler of Egypt who was buried within. The Khafre Pyramid was built next, followed by the smallest of the group, the Menkaure Pyramid. When they were first built, each had an outer layer of white limestone and was blindingly white in the desert sun. Today the metropolitan area of Cairo reaches to the foot of the pyramids. This image was collected by the QuickBird satellite on February 2, 2002.

161

Intensive Agriculture

POLAND

Humans have an unmatched ability to alter the landscape. This image shows an area just southwest of the city of Lublin, Poland, near the Chodel River. The patchwork of fields indicates intensive agricultural activity. Thousands of years ago the area was completely covered by dense forests, but today only a few small patches remain. This image was acquired by the EROS A satellite on June 20, 2002.

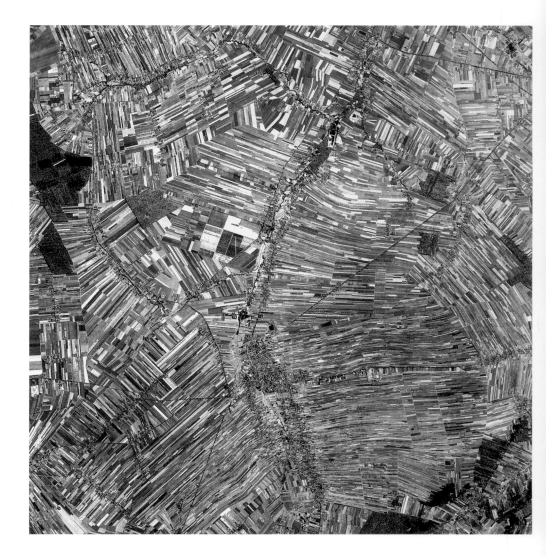

Changing the Way We Look at the World

One of our most important attributes as human beings is the ability to express ourselves artistically. People are inspired to create art as a response to the world around them, not only reflecting realistic scenes, but also conveying emotions, ideas and states of mind.

The advent of remote sensing has had an important effect on the art world. From the Renaissance until the 19th century, most artists attempted to represent the world around them accurately. This approach began to change in the 1870s. Photography not only made it unnecessary for artists to represent things faithfully, but it also challenged them to understand and depict the world around them differently. Artists began to experiment with ways of representing three dimensions on a two-dimensional canvas, and they often fragmented or flattened their images.

In 1899 the Eiffel Tower opened in Paris, the center of the art world at the time. It was the first time the general public could obtain a perspective of the Earth's surface from so high above. Art critics have noted that the view of the ground from the top of the tower was more important than the view of the tower on the skyline. Paris, from a thousand feet up, now became its own map with different patterns extending to the horizon. Poets wrote of the Eiffel Tower overcoming the bounds of Earth, and artists were moved to include this type of perspective in their art.

The first airplanes, widely demonstrated by the Wright brothers in Europe, revolutionized the way people imagined the world around them. In fact the artist Pablo Picasso often referred to Georges Braque, his fellow developer of the Cubist movement, as "Wilbourg," taking the nickname from aviation pioneer Wilbur Wright. Cubism, like an overhead view of Paris, emphasized flat two-dimensional patterns broken by linear features while simultaneously representing the three-dimensional world. It was a way of depicting the world freed from the bounds of the canvas, just as the airplane freed humans from the bounds of the Earth's surface. Contemporary artists continue to be inspired by remote sensing images of Earth.

Human Population

Before industrial societies developed in the 1800s, population growth was slow. In 1900, the Earth's population was about 1.5 billion. During the 20th century the population grew by almost 2 percent a year, almost doubling every 35 years, until by 2000 it had reached 6 billion.

Better health care and sanitation have increased life spans. As life expectancies increased, the population exploded. After several generations, living standards also rose, causing birth rates to fall as the population changed its focus. Many North American and European nations, as well as Japan, followed this pattern, growing explosively in the 18th–20 centuries. Population growth has dramatically slowed in many countries, and by 2000 some European nations began to experience population decline. At the same time the nations in South America, Africa and southern Asia continue to experience huge population growth. As economic growth increases living standards in these nations, population growth may slow. In 1950 the average woman in a developing nation gave birth to about six children, but by 2000 the number had dropped to about three. If these trends continue, the world population will stabilize at about 11 billion people by 2100.

Growing Our Food

No human activity is more widespread on the Earth than agriculture, which covers about 15 percent of the Earth's land surface. Crops remove the natural vegetation cover, and irrigation changes the flow of water over the surface. About one-third of the world's population is involved in agriculture.

Intensive use of the land can lead to serious problems. In the 1920s drought and overly intensive agricultural practices led to the Dust Bowl, a time when dry topsoil was eroded by wind on the plains of the central United States. Today, some of the most severe erosion is taking place in northern China, where more than 100 tons of soil per acre are eroded into the Huang He (Yellow River) every year. Agriculture has also displaced many species of plants and animals. There are millions of natural species on Earth, but only about 200 plant and 30 animal species are used in agriculture. The widespread extent of farming may be causing dramatic decreases in global biodiversity.

Satellite images are used to forecast and improve crop yields, and agricultural applications provided one of the initial justifications for satellite remote sensing. Because vegetation reflects very strongly in near-infrared wavelengths, satellites can directly detect the presence and density of plants on the ground. And since crops are planted in very regular rows and at specific times, scientists can accurately determine the status of crops on the ground and estimate crop yields.

Urban Landscapes

Human society is rapidly becoming more urbanized. In 2000 about half the Earth's population lived in urban areas, and that figure may reach 60 percent by 2020. Satellite images show the growth of these areas and how they change. During the 19th and 20th centuries most large cities developed suburban areas surrounding the city centers. Suburbs began as rail transportation reached outlying areas, allowing people to live outside the city and commute into town for work.

Following World War II, the United States led the way in expanding the suburban landscape. By 1955 three of every four Americans owned cars, and these suburban areas grew dramatically. In 1956 the United States began building the interstate highway system, which would eventually span the country. Between 1980 and 2000, the population of the suburbs surrounding the largest American cities grew by an average of 44 percent.

Because buildings and pavement reflect sunlight differently than natural surfaces do, remote sensing data can be used to produce maps that show city planners and decision makers where development is occurring. These maps can be updated more often than traditional ones, allowing for quick responses.

Satellites can also detect the heat generated by cities. Urban surfaces such

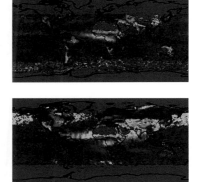

Smoke and Haze in the Atmosphere

These global maps show the extent of smoke and aerosols in the atmosphere. Although all aerosol particles are microscopic, those that occur naturally tend to be larger than those produced synthetically. Orbiting sensors can estimate the size of the particles floating in the atmosphere, allowing scientists to discriminate between human-produced and natural particles. In this map natural aerosols are green, pollution is red and a mixture of both is shown in light brown. The maps were produced from data from the MODIS sensor on the Terra satellite. These two maps were made six months apart.

as pavement and buildings absorb and trap more heat from the sun than natural surfaces do. Concentrations of buildings and vehicles produce their own heat, creating an area of high temperature called the urban heat island. Temperature differences of about 9°F (5°C) are common between a central city and outlying suburbs.

Pollution

Pollution results from the release of human-produced materials that cannot be readily recycled by the Earth's biological or environmental processes. It disrupts ecosystems and spoils landscapes. Agriculture is a primary source of water pollution. Nitrates and phosphates, used as fertilizers, can cause extensive damage to lakes and rivers. The burning of fossil fuels adds gases and other materials to the air, affecting breathing over wide areas. More than 40 percent of oil pollution in the seas comes from industrial runoff. Spills from oil tankers are dramatic events, but they account for only about 20 percent of the oil released into the oceans. Far more comes from poorly maintained engines of water craft. Advances in shipping technology have decreased oil spills 75 percent since 1970. The presence of oil in the oceans changes the way waves form on the water's surface. Radar sensors in orbit can be used to map where these changes take place, allowing oil recovery teams to be positioned where they are most needed.

Every day about eight million pieces of trash accidentally fall from cargo ships into the Earth's oceans. Mostly made of plastic, they are carried by ocean currents far from their places of origin. Thousands of pieces of floating trash have been found on uninhabited Pacific islands. Every year about 10,000 shipping containers, about 0.01 percent of all those carried on cargo ships, are swept overboard into the ocean.

Treating sewage is a challenge for modern societies. Most large North American and European cities spend great amounts of money to filter and clean waste materials before releasing them. However, this is not always the case. Until 1992 the metropolitan area around New York City dumped about eight million tons of sewage into the Atlantic Ocean every year. Nations without the resources to devote to sewage treatment usually release wastes into rivers and oceans.

Air pollution, when it collects in large amounts, obscures the Sun. The burning of fuels consumes oxygen from the atmosphere and releases carbon dioxide and microscopic particles that float in the air, giving the sky a hazy appearance and causing irritation to eyes and lungs. Photochemical smog is created when exhaust reacts to sunlight and produces a thick, brownish haze

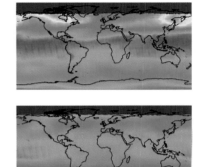

Ozone

Ozone in the upper atmosphere protects the Earth from harmful solar radiation. Beginning in the 1920s, chemicals called chlorofluorocarbons (CFCs) were widely used in industry and as propellants in products such as hair spray. The production of chlorofluorocarbons severely lowered the amount of ozone in the atmosphere. Their use has now been restricted, and the ozone layer is recovering. However, the recovery is not complete, and every winter the amount of ozone is still very limited over Antarctica. These maps were made six months apart using data from the Total Ozone Mapping Spectrometer instruments on several satellites. Dark colors indicate low concentrations of ozone in the upper atmosphere and bright colors indicate high concentrations.

in the atmosphere. Satellite sensors can detect this haze from above. Layers of pollution become particularly visible when a temperature inversion takes place in the atmosphere, with warm air high in the atmosphere trapping cooler, polluted air close to the surface.

Deforestation

Deforestation has led to the loss of much of the forest cover that has existed for thousands of years. It takes place when people migrate to forested areas and clear the land for agriculture, or when they harvest trees for fuel or lumber. This leaves a pattern of cleared fields within the forests. Much of Europe and eastern North America lost their natural forests as people cleared the land to build thriving economic systems. During the 20th century this trend partly reversed, as many marginal agricultural lands in North America reverted to forest cover.

In Asia and South America, however, the trend is toward losing forest cover. Roads are built across vast forests to accommodate settlers and loggers. It is likely that about 10 percent of the tropical forests once standing in Brazil were removed by the year 2000. Population increases and unfavorable economic conditions on the Atlantic coast have encouraged Brazilians to move to the interior of the country. In 2003 about 160,000 families lived in squatter camps in the interior, nearly 100,000 of them arriving during that year alone.

The patterns of deforestation are clearly visible from orbit. Governments make maps of landcover change to define where forests are being cleared. These maps can have an important effect on development schemes. The first estimates of deforestation in South America showed alarming rates of forest loss. Although some more recent estimates showed lower rates, vast amounts of land are still being altered. Global estimates are that about 60,000 square miles (150,000 sq km) of forest are lost every year, about 7,500 square miles (20,000 sq km) in Brazil alone.

Desertification

Desertification is the human-induced change of landcover from non-desert to desert. This can happen when people remove much of the vegetation from the land in arid areas.

Much of the Earth is covered in drylands, which include the savannas and plains where millions of people make their homes. Drylands are also home to the Earth's most productive farmlands and are where most of the world's grain production takes place. The majority of the Earth's drylands are threatened by desertification.

Scientists have searched for signs of desertification in the Sahel region of

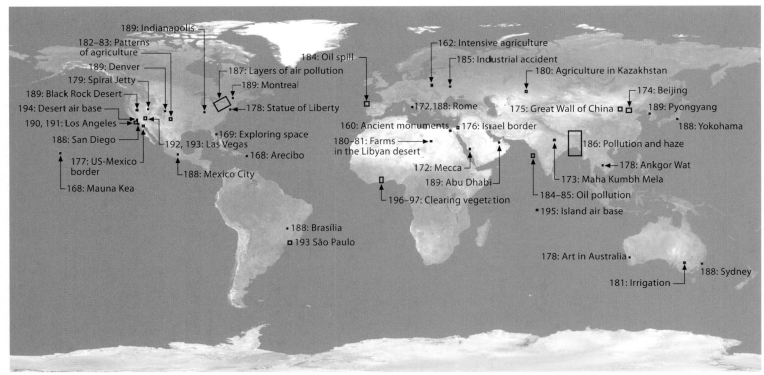

Numbers indicate page locations of the images in this chapter.

Africa. The Sahel stretches across the western part of the continent at the southern edge of the Sahara Desert. It is a transition zone between the hyper-arid desert to the north and the rainy forests to the south. Since first being mapped using satellite imagery in the 1970s and 1980s, the desert boundary has regularly undergone movement north and south in response to shifts in climate.

Borders between Nations

Most of the maps and globes used to represent the Earth are political maps, which show nations and their borders. Physical maps, used by scientists and engineers, show features like rivers and mountains. Most maps show a combination of both political and physical features. Astronauts in space frequently comment that, as they look down at the Earth, it is not possible to see political borders from above.

Still, some nations do have borders that are visible from space. Often two neighboring nations use the land differently. There may be two adjacent areas with different patterns of irrigation that distinguish the vegetation on the ground. If populated areas from two nations are next to each other, it is often easy to see different development patterns from above. Sometimes countries are not good neighbors, and they build structures to prevent military incursions. These features can include fences or ditches, and they can sometimes be large enough to change the appearance of the land.

The World's Largest Telescopes

ARECIBO RADIO TELESCOPE (ABOVE)
Just as astronomers use telescopes to explore the universe visually, they use radio telescopes to listen to faint emissions from stars and other objects in space. This image, acquired by the IKONOS satellite on January 7, 2002, shows the Arecibo Radio Telescope in Puerto Rico. Built in a natural depression, the 900 foot (300 m) wide dish antenna is the largest in the world. The detector is suspended 550 feet (170 m) above the reflector. Arecibo can also transmit and detect radar waves, a capability that scientists have used to map features on the Moon and several planets.

OBSERVATORIES ON MAUNA KEA, HAWAII (RIGHT)
The peak of Mauna Kea, an extinct volcano in Hawaii, is home to the world's largest telescopes. Its high altitude, at 13,796 feet (4,205 m), decreases atmospheric interference, and its location near the equator allows good visibility over much of the sky. This image, acquired by the IKONOS satellite on January 2, 2003, features at least eight observatories, including the Subaru telescope and the twin domes of the Keck Observatory, the largest optical telescopes in the world.

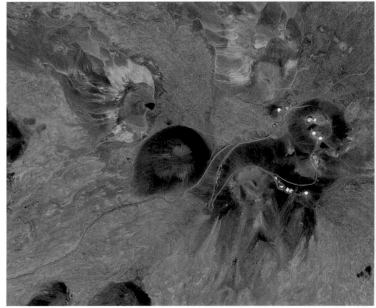

Exploring Space

KENNEDY SPACE CENTER (ABOVE AND LEFT)

The Kennedy Space Center in Florida is a launch site for U.S. space exploration vehicles. The top image, collected by the IKONOS satellite, shows the 525 foot (160 m) high Vehicle Assembly Building, one of the largest enclosed spaces in the world. During the Apollo program in the 1960s, the Saturn rockets that carried astronauts to the Moon were assembled here. Today the Space Shuttle, in preparation for launch, is mated with its external fuel tank and rocket boosters inside the building.

The image on the left, acquired by the QuickBird satellite, shows one of two Space Shuttle launching pads. Built to launch Saturn rockets to the Moon, they were modified to accommodate the Space Shuttle, which was first launched in 1981. The tower is 346 feet (105 m) tall to the top of its lightning rod. The blast from the rockets is deflected through openings in the center of the pad. In the upper part of the image is a water tank about 300 feet (90 m) high. Water from the tank is dumped onto the pad during launches to absorb the violent sonic energy.

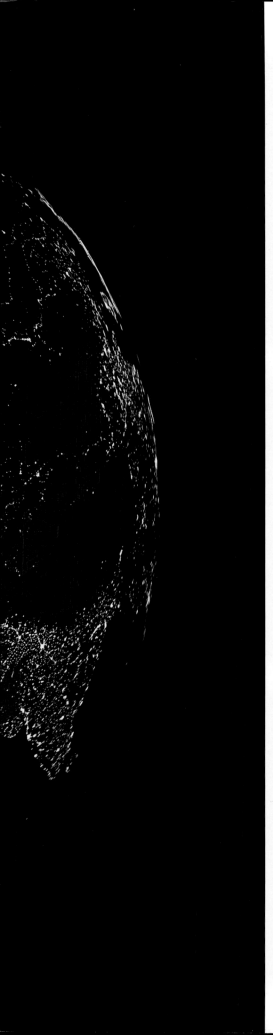

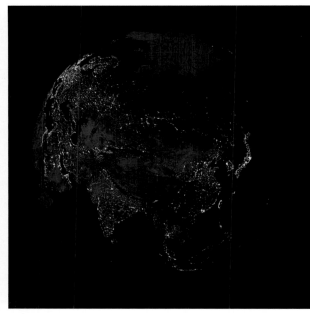

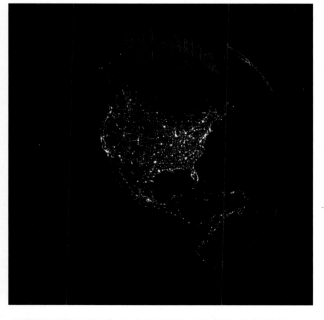

Earth at Night

These global maps show the Earth at night, when the human presence on the globe is most visible through our use of artificial light. Streetlights are the source of most of the light seen in these images. Areas with little population, such as the deserts of northern Africa and central Asia, are devoid of light. Densely forested places, such as central South America, are also dark. In the United States the paths of highways can be discerned. European cities are especially bright. The route of the Trans-Siberian Railroad has created a line of well-lit cities strung out across Asia. The course of the Nile in Egypt is visible. These maps were made using data from a sensor on the DMSP satellites. A dark-blue layer was added to show where land masses are located.

Holy Pilgrimages

VATICAN CITY (ABOVE)

MECCA, SAUDI ARABIA (RIGHT)

Nothing has a more powerful effect on human civilization than religious faith. The image on the right, acquired by the QuickBird satellite on February 11, 2003, shows Mecca, Saudi Arabia. Mecca was the birthplace of the prophet Muhammad in 570 AD. Today millions of people travel there to perform the Hajj, the pilgrimage to the holiest places in Islam. The al Haram mosque, which can accommodate more than 400,000 worshippers, appears near the center of the image. At its center is the Ka'bah, a cube-shaped shrine that

tradition states was built by Abraham and his son Ishmael.

The image above, collected by the IKONOS satellite on May 5, 2003, shows Vatican City. Occupying about 108 acres (44 hectares) within Rome, Italy, it is the world's smallest independent state. The Vatican is the administrative and spiritual center for the Catholic Church. The buildings of Vatican City are dominated by Saint Peter's Basilica, the large domed building at the center of the image. People travel from around the world for services and special events in Saint Peter's Square (actually an oval-shaped plaza) visible to the right of the basilica.

172

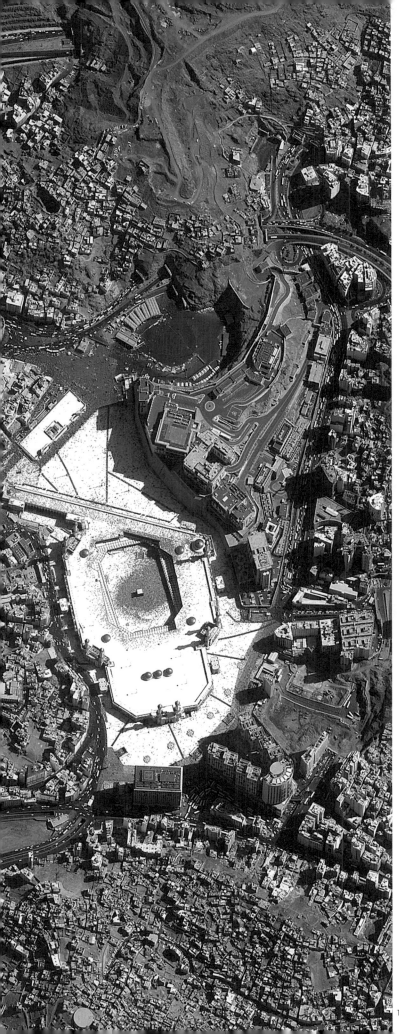

Maha Kumbh Mela

GANGES RIVER, INDIA (ABOVE)

This image shows a Hindu pilgrimage on the Ganges River in India. During the Maha Kumbh Mela festival, faithful Hindus gather at the spot where the Ganges River meets the Yamuna and the mythical Saraswati rivers near the city of Allahabad. As part of the ritual, many people bathe in the river. The pilgrimage has taken place every 12 years in Allahabad since at least 500 AD.

The 2001 Maha Kumbh Mela was seen as a particularity auspicious time for the pilgrimage. Attracting tens of millions, it was probably the largest gathering of humans in the history of the world. The IKONOS satellite acquired this high-resolution image on January 23, 2001, the most sacred day of the festival, when about 30 million people bathed in the river. Temporary bridges cross the river, and a city of tents has been built on the right. Crowds of people are visible in the water.

Marks of Civilization

BEIJING (LEFT)

The high resolution image on the left was acquired by the QuickBird satellite on February 11, 2002. It shows the Forbidden City at the very center of Beijing. This area, the center of Chinese political power from 1420 to 1911, was protected by walls and a wide moat. Common people were not allowed to enter the Forbidden City, giving the area its name. To the south (on the right side of this image) is Tiananmen Square. One of the largest public spaces in the world, it can hold a million people.

GREAT WALL OF CHINA (LOWER RIGHT)

The Great Wall of China was constructed to prevent invasions from the north. Early barriers to invasion were built in central China as early as 650 BC. Under the Ming Dynasty in the 1400s, China made it a priority to strengthen the northern border. Most of the wall that is still visible today was built during this period. The most famous section of the wall, within 60 miles (100 km) of Beijing, was restored in the 1950s.

The wall reaches from the Gobi Desert to the Korean Peninsula, a distance of 4,500 miles (7,240 km). Its average height exceeds 20 feet (6 m). The western portion of the wall is made of compacted dirt, but in the east the wall structure is far more complex.

A black line running diagonally through the image shows the Great Wall in northern Shanxi Province. The fresh snow cover on the land makes the wall more distinctive than it would normally be from space. This image was recorded by the ASTER sensor on the Terra satellite on January 9, 2001.

ANGKOR WAT (TOP RIGHT)

This IKONOS image from April 12, 2004 shows Angkor Wat in northwestern Cambodia. Angkor Wat was built by the Khmer Empire in the early 1100s as a Hindu temple. It was later converted for use as a Buddhist temple. The temple buildings are visible in the center of the image. Surrounding that is a moat 630 feet (190 m) wide that encloses an area of 203 acres (83 hectares). The green color indicates forest vegetation.

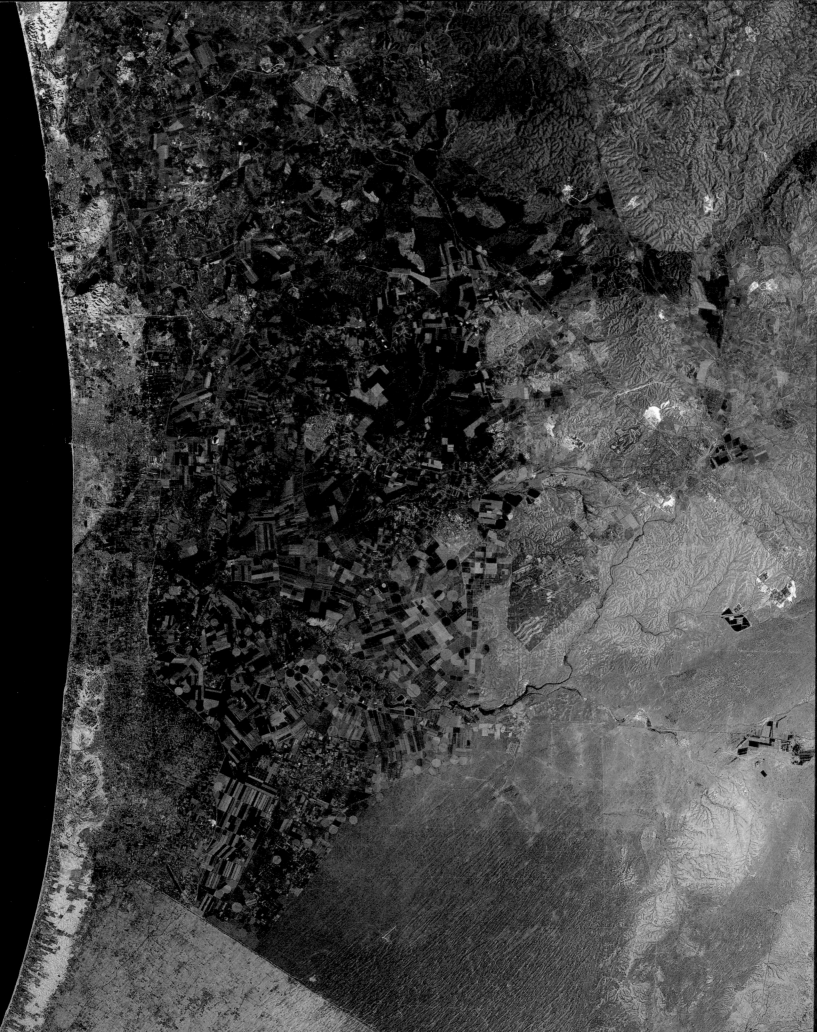

Visible Borders

ISRAEL–GAZA (LEFT)

Political borders are not often visible from space, but Israel's borders with Egypt and the Gaza Strip can be seen in this Landsat image. Differences in land use and irrigation practices make the border with Egypt plainly visible from space.

The Gaza Strip is about 25 miles (40 km) long and, at its widest point, about 5 miles (8 km) wide. Beginning in 1994 much of the territory was administered by the Palestinian Authority. The border with Israel is fortified with fences and walls. Cleared land along the margins of the Gaza Strip makes the border visible from space.

UNITED STATES–MEXICO (ABOVE)

The U.S.–Mexico border is the straight line running through the center of this image. Mexicali, Mexico, is south of the border next to Calexico, California. El Centro, California, is in the upper left.

Agricultural fields appear bright red in this near-infrared image, collected by the ASTER sensor on the Terra satellite on June 12, 2000. More than 3,000 miles (5,000 km) of irrigation channels deliver water to the fields. In 1901 the Imperial Canal was opened, diverting water from the Colorado River. Four years later floods overwhelmed the irrigation channels. The excess water settled in a natural depression to the north, creating a huge brackish pool called the Salton Sea. During the 1930s the All-American Canal was built, diverting still more water from the Colorado River. Fields on the Mexican side are not as intensely irrigated, and competition for water is often a source of friction between the two countries.

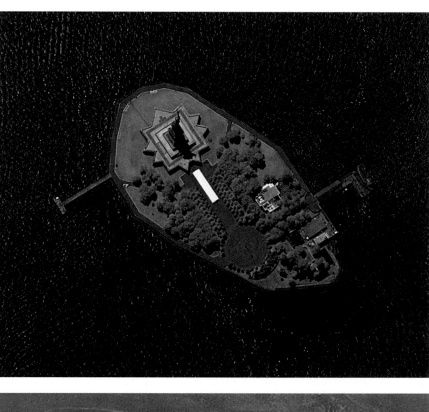

Human Creativity

STATUE OF LIBERTY (TOP LEFT)
The Statue of Liberty in New York harbor was a gift from France to the United States. The statue is 151 feet (46 m) tall and is made of 225 tons of copper and iron. Construction began in 1875 under the direction of sculptor Frédéric-Auguste Bartholdi, and the statue was shipped to Bedloe's Island (now called Liberty Island) in 1885. Originally the statue's torch was used by the the U.S. Lighthouse Service for navigation. In the 1980s the statue was restored, and a new gold-covered torch replaced the original. This image was collected by the IKONOS satellite on June 30, 2000.

PEACE IN AUSTRALIA (LOWER LEFT)
This IKONOS satellite image shows a large-scale artwork near New Norcia, Western Australia on January 26, 2007. Titled *Peace*, it was created by arranging colored sand, gravel and plastic within a 100 meter-wide enclosure on farmland. It was unveiled on Australia Day to mark the 40th anniversary of a 1967 referendum that recognized Aboriginal people as citizens. *Peace* was created by three artists of Yuat Aboriginal heritage: Sheila Humphries, Fatima Drayton, and Deborah Nannup. The artwork represents six campsites and a central meeting place near the Moore River.

SPIRAL JETTY (RIGHT)
In the 1960s artists began to create monumental pieces in natural settings, often called "environmental sculptures." In 1970 Robert Smithson used basaltic rocks and dirt to create the Spiral Jetty in Great Salt Lake, Utah. The earthwork is 1,500 feet (450 m) long by 15 feet (4.5 m) wide. The long, straight structure nearby was part of an oil-drilling operation in the 1920s through the 1980s. The jetty, often submerged, was clearly visible in this IKONOS image from September 2002, when Great Salt Lake was experiencing low water levels. More images of the lake can be seen on pages 244 and 245.

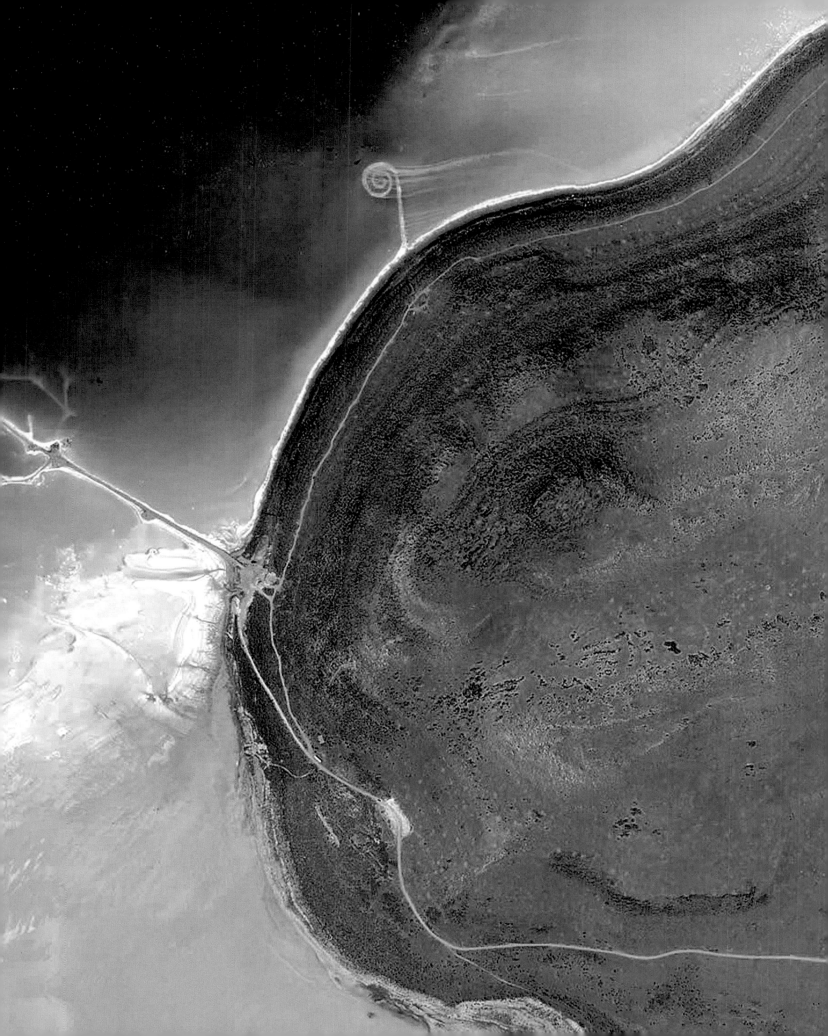

Patterns of Agriculture

KAZAKHSTAN (ABOVE)

This photograph, taken from the Space Shuttle in October 2002, shows land near the city of Arkalyk in central Kazakhstan. At the right side of the image, large fields of wheat are being cultivated. Patterns on the left side show bauxite and asbestos being mined directly from an open pit. Both mining and agriculture require the intensive use of water in this arid area.

LIBYA (RIGHT)

It is possible to grow crops in environments as dry as the Sahara Desert. The center image shows how irrigation is used at the Al Kufrah Oasis in southeastern Libya. Water extracted from aquifers deep beneath the desert is used for center-pivot irrigation. In this common method of watering crops, water is sprayed from a long arm that moves in a circular motion, creating distinctive circular patterns. This image was acquired by the EROS A satellite in December 2002.

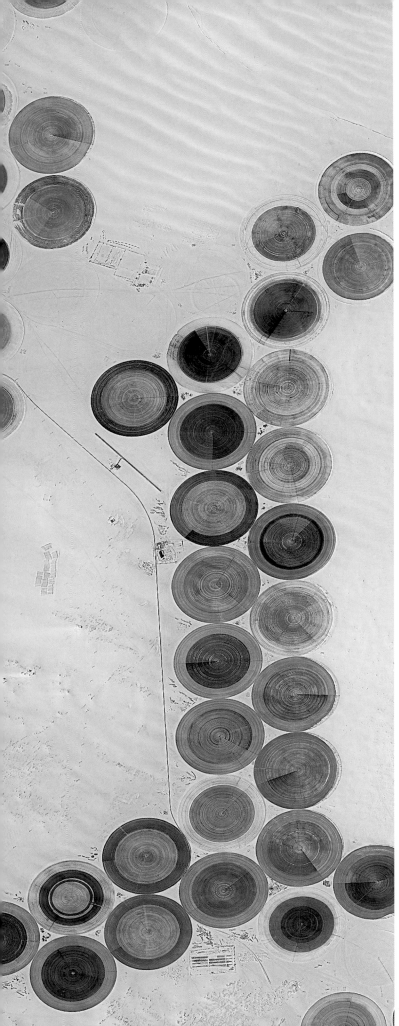

AUSTRALIA (ABOVE)

Irrigation is also required to make agriculture successful in dry Australia. This photograph was taken from the International Space Station on November 21, 2002. It shows agriculture at Lake Tandou in the Menindee Lakes region of New South Wales. The Darling River flows past the lake to join the Murray River. Canals up and down the river connect Lake Tandou to other lakes; and, when needed for irrigation, water is released from the lakes. This area is always dry, but was experiencing severe drought when the photograph was taken in November 2002. Only 10 inches (25 cm) of rain had fallen in two years, making it the worst drought in a century.

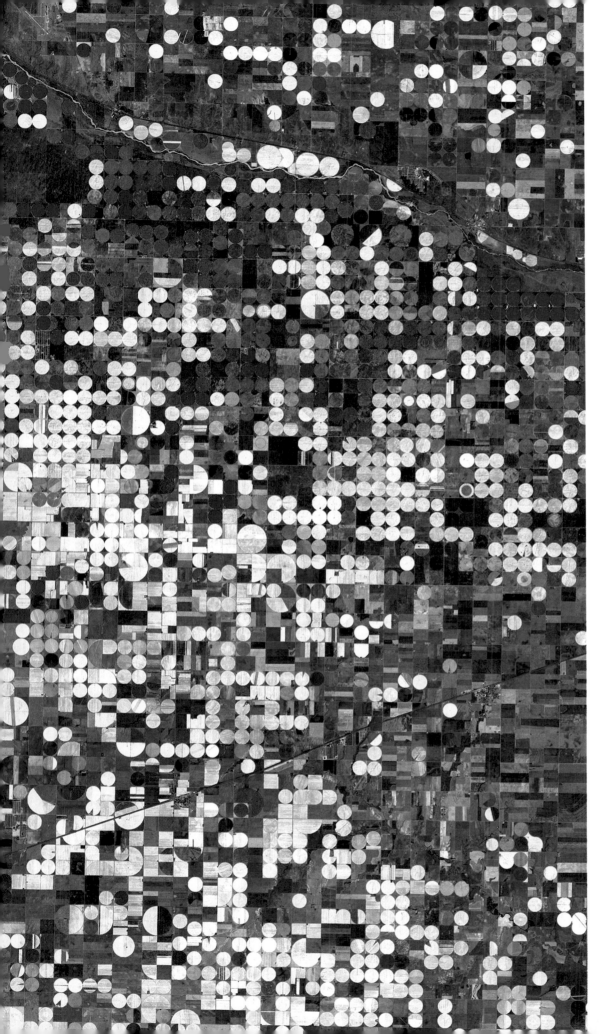

Patterns of Agriculture
GARDEN CITY, KANSAS

Agriculture is the most widespread use of land by humans. This image, acquired by the Landsat 7 satellite on September 25, 2000, shows hundreds of irrigated fields in the area south of Garden City, Kansas. More wheat is grown in Kansas than in any other U.S. state. Crops that reflect near-infrared wavelengths appear bright red. Light-colored fields are fallow or have already been harvested. The Arkansas River flows through the northern part of the image.

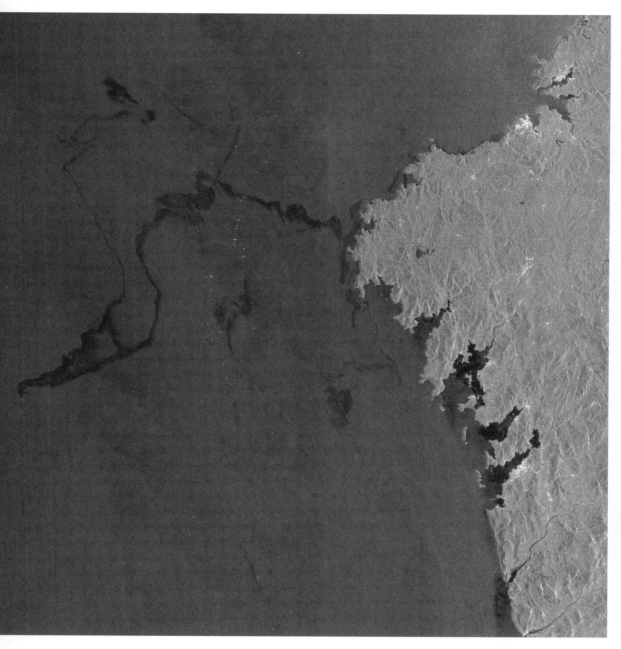

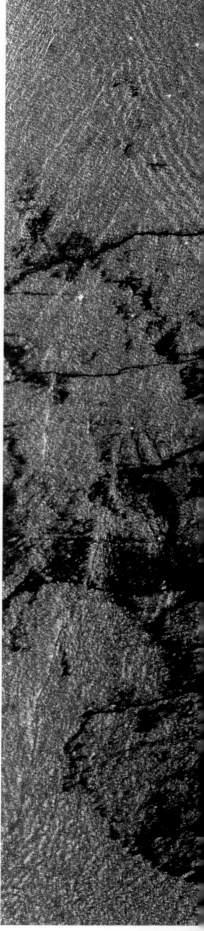

Oil Spills

OFF THE COAST OF SPAIN (ABOVE)
ARABIAN SEA (RIGHT)

Petroleum is the fuel that drives modern economies, but accidents can occur. Oil spills can happen at drilling sites or along the shipping routes used to transport petroleum to markets. In November 2002 the tanker Prestige was rounding the Galician coast of northwest Spain. After a violent storm, a crack formed in the hull and the tanker began leaking oil. Over the next six days the tanker drifted out to sea, eventually breaking in half and sinking on

November 19. About 3 million gallons of oil were released, but the ship took many millions more with it to the bottom of the sea. The radar image above, collected by the ASAR sensor on the Envisat satellite on November 17, shows the Prestige as a white dot about 60 miles (100 km) off the coast. The oil slick, which can be seen trailing behind the ship, has already reached the coast. Support ships can be seen between the ship and the coast.

The radar image on the right, from the SIR-C Space Shuttle mission in October 1994, shows a field of drilling

platforms in the Arabian Sea, about 150 km (93 miles) off the coast of Mumbai, India. Drilling platforms are visible as white dots. Oil slicks, shown in black, eventually spread over large areas. The oil smoothes waves, making the slicks visible in radar images. Surface winds create the small ocean swells that are seen throughout the image. Color differences in the image indicate small waves and swells created by surface winds.

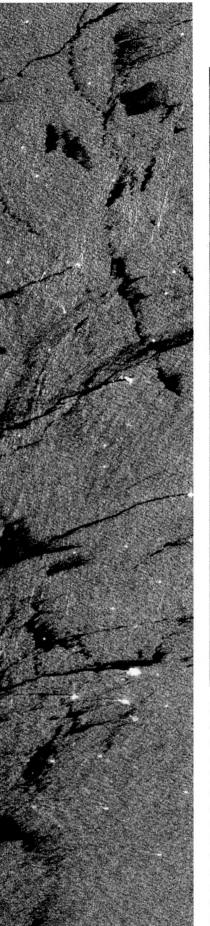

Industrial Accident

CHERNOBYL, UKRAINE (ABOVE)

On April 26, 1986, the power level at one of the nuclear reactors of the Chernobyl power plant in Ukraine was allowed to drop below safe levels in a planned experiment that went terribly wrong. The uranium core melted and caused two large explosions. The top of the building was destroyed and, for the next 10 days, radioactive material was carried over much of Europe. Some 100,000 people were evacuated, and about 30 were killed by exposure to radiation while cleaning up the site.

This image, collected by the EROS A satellite on April 14, 2003, shows Chernobyl 17 years after the accident. The nearby city is still deserted. Large cracks have been seen in the huge concrete barrier (nicknamed the sarcophagus) constructed to contain the wrecked building.

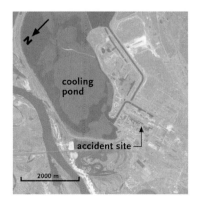

cooling pond

accident site

2000 m

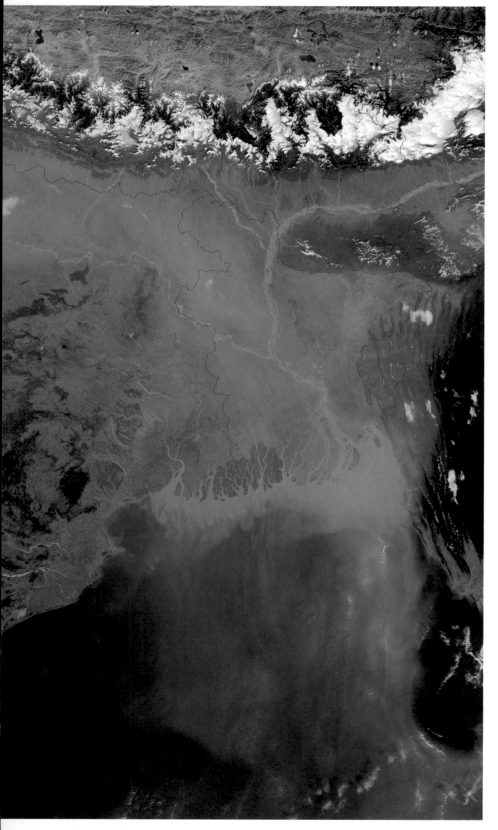

Pollution and Haze

BANGLADESH (LEFT)

The image on the left shows thick air pollution over Bangladesh. The area around Bangladesh and eastern India is one of the most densely populated regions on Earth. Heavy pollution is generated in Calcutta, Dhaka and other cities. This image was captured by the MODIS sensor on the Terra satellite on December 4, 2001, when a mass of warm air arrived over slightly cooler air. This temperature inversion prevented pollution from rising in the atmosphere. The Himalayan Mountains to the north trapped the pollution until winds picked up and cleared the air.

Layers of Air Pollution

THE GREAT LAKES (RIGHT)

A high layer of clouds is seen in the upper part of the photograph to the right. Below that is a thick layer of pollution mixed with water vapor. This photograph was taken by Space Shuttle astronauts on October 21, 2000. Sunlight is reflected in Lake Ontario (foreground), Lake Erie (background) and the Finger Lakes. The city of Buffalo is located between Lake Ontario and Lake Erie, and Rochester is located in the center of the photograph on the south shore of Lake Ontario.

Great Gathering Places

Cities all over the world have huge stadiums for sporting events, concerts and festivals. These satellite images show gathering places in different continents.

The International Stadium in Yokohama, Japan, with a capacity of 72,000, was the site of the final match of the 2002 World Cup. The Mané Garrincha Stadium in Brasília, Brazil, is shown in a QuickBird image. The Palacio de los Deportes is a 21,000-seat indoor arena in Mexico City. The famous Coliseum in Rome, completed in 80 AD, has room for more than 50,000 people. Some of the original structure under the main floor is visible in the center of the building. Qualcomm Stadium in San Diego is home to professional sorts teams.

The Burning Man festival takes place annually in the Black Rock Desert near Gerlach, Nevada. The IKONOS image from September 3, 2005 on the opposite page shows around 30,000 people camped there for the alternative arts and lifestyle festival. Invesco Field and Mile High Stadium in Denver are shown in an IKONOS image from August 2001. Mile High Stadium has since been torn down.

Olympic Stadium in Montreal was built for the 1976 Olympic games. The tower was completed in 1987. The Indianapolis Raceway is home of the annual Indianapolis 500. Sheikh Zayed Sport City in Abu Dhabi, United Arab Emirates, has room for 60,000 people. May Day Stadium lies on a island in the Taedong River in Pyongyang, North Korea. This is the largest stadium in the world, with seating for 150,000.

▲ YOKOHAMA

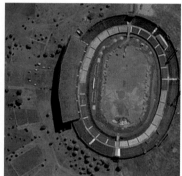

▲ BRASÍLIA

▲ MEXICO CITY   ▼ ROME

▲ SAN DIEGO   ▼ SYDNEY

▲ NEVADA

▲ MONTREAL

▲ INDIANAPOLIS

▲ ABU DHABI

▼ DENVER

PYONGYANG ▼

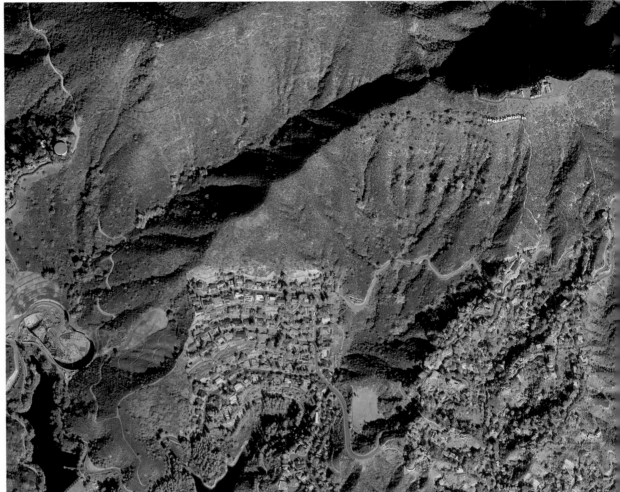

Sprawling Cities

LOS ANGELES (LEFT)

Los Angeles is the second-largest city in the United States and its growth is evidence of the power of humans to alter the landscape. In 1822, when Los Angeles became part of an independent Mexico, about 1,000 people lived in the city. Between 1900 and 1910, as part of the United States, Los Angeles' population tripled to over 300,000. Sitting in a desert, it did not have a reliable source of drinking water and lacked a natural harbor. These problems did not slow the growth of Los Angeles for long. To solve the water problem the first of several aqueducts was completed in 1913, carrying water from the Owens Valley 250 miles (420 km) to the north. The next year an artificial harbor opened. With the completion of the Panama Canal, Los Angeles quickly became the busiest port on the Pacific Coast and remains one of the busiest ports in the world. In 2000 the population of metropolitan Los Angeles was over 10 million. This image was produced by combining data from the Landsat 7 satellite and the Space Shuttle Topography Mission.

HOLLYWOOD (ABOVE)

Hollywood, about 8 miles (14 km) northwest of downtown Los Angeles, was laid out in 1887 and became absorbed into the city of Los Angeles in 1910. The area's first movie studio opened in 1911, and Hollywood soon became synonymous with the film industry. The famous Hollywood sign is visible in the image, acquired by the IKONOS satellite. The sign was built in 1924 — originally, with the spelling "Hollywoodland" — as an advertisement for a new housing development. The development never came about, but the sign was refurbished and shortened.

191

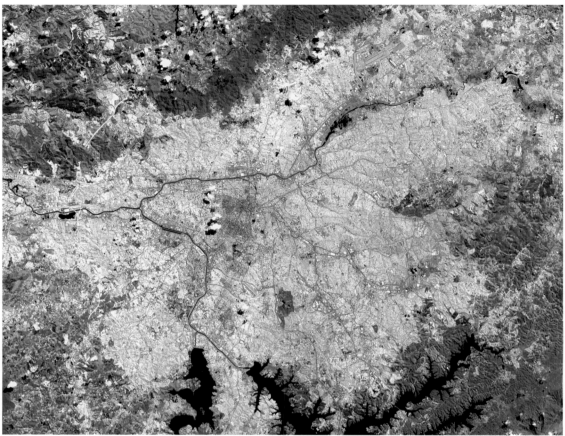

Sprawling Cities

LAS VEGAS (LEFT)

Las Vegas, Nevada, began as a small desert crossroads. The Hoover Dam was built on the nearby Colorado River in the 1930s, and gambling, which attracted tourists, was legalized in 1931. Casinos and resorts are still the backbone of the economy, and the center of the casino business is the "Strip," which lies between McCarran International Airport and downtown. Las Vegas is one of the fastest-growing cities in the United States. The metropolitan area's population doubled to over one million between 1985 and 1995, and the area covered by development also doubled during those years. This image from October 17, 2000, acquired by the ASTER sensor on the Terra satellite, shows new streets being prepared around the city's edges.

LAS VEGAS AT NIGHT (ABOVE)

The ASTER sensor on the Terra satellite acquired this nighttime image of Las Vegas on June 12, 2000. It shows the city in thermal-infrared light. The brighter areas indicate higher surface temperatures. Urban surfaces can change the temperature of their surroundings. Buildings and streets absorb and retain more energy from the Sun than natural surfaces do, creating higher temperatures in urban areas as the buildings and streets release their heat.

SÃO PAULO (ABOVE)

São Paulo, Brazil, has sprawled across vast areas and is the largest city in South America. In this image from the ASTER sensor on the Terra satellite, acquired on March 19, 2002, green indicates grasses, forest cover and other types of vegetation. Urban areas appear in shades of gray and blue.

In 1554 Catholic missionaries from Spain founded São Paulo on the banks of the Tietê River. Today, more Roman Catholics live in São Paulo than in any other city in the world. In 1870 São Paulo was one-tenth the size of Rio de Janeiro, but coffee cultivation and then industrialization led to explosive growth. During the late 1800s immigrants flocked to São Paulo from Europe and Asia. São Paulo is home to more people of Japanese descent than any city outside of Japan. By 1970 São Paulo was Brazil's largest city, and every year another 300,000 people arrive from around around the country seeking economic opportunities.

Desert Air Base

EDWARDS AIR FORCE BASE, CALIFORNIA (LEFT)

Evidence of human activity can be seen even in remote places. Edwards Air Force Base in California was built in the Mojave Desert northeast of Los Angeles. The site was chosen because it is a dry lake bed and provided ample room for takeoff and landing. Runways are indicated by lines painted on the desert floor. This image from the IKONOS satellite also shows a huge compass drawn on the ground to guide pilots. The facilities at Edwards are used to test new aircraft. Among the many aircraft visible here are a B-52 bomber and two SR-71 reconnaissance aircraft in the paved area near the lower-left corner.

Island Air Base

DIEGO GARCIA, INDIAN OCEAN (RIGHT)

On the other side of the Earth from California is Diego Garcia, an island in the middle of the Indian Ocean. In 1965 Diego Garcia became part of the British Indian Ocean Territory, and the next year an agreement was signed with the United States allowing the construction of an airbase on the island. The small population was resettled on the island of Mauritius, 1,500 miles (2,400 km) to the southwest. Today the island is still British territory, but it is almost completely covered by the U.S. Air Force base. This high-resolution image from the IKONOS satellite, July 10, 2000, shows runways and buildings on the island.

Clearing Vegetation

GHANA (RIGHT)

This image, acquired by the Landsat 7 satellite on February 2, 2002, shows cleared vegetation and deforestation in southwestern Ghana and southeastern Côte d'Ivoire. The ocean is visible to the left (north is to the right). The darker areas indicate denser standing vegetation. The border between the two nations runs along the upper part of the image, and Abidjan, the capital and largest city of Côte d'Ivoire, is visible in the upper-left corner.

The large geometric shapes show the extent of standing forests. The borders have been defined by property boundaries and political units. Agriculture has long been the economic basis for most people in Ghana. Cocoa was introduced in the 1500s, and the native vegetation was intensely cleared during the 1800s. Today Ghana produces more cocoa than any other nation.

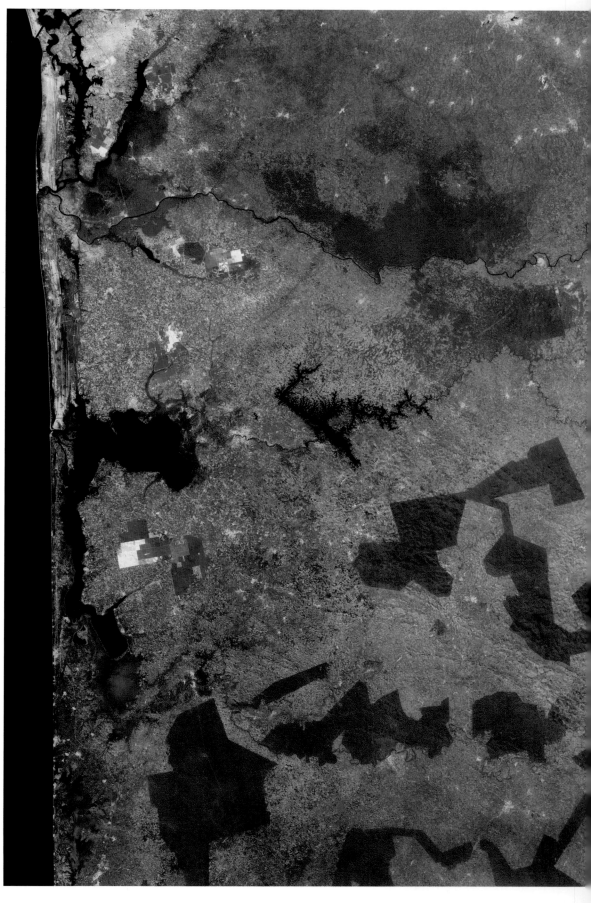

Roads of Commerce

Humans have been traveling across the surface of the Earth for thousands of years, trading goods in an effort to build better lives. The structures that accommodate travel and trade, such as cities and highways, are plainly visible from space. In the history of building societies and governing them, there have of course been different motivations and practices. One longstanding theme, however, is the search for trading opportunities. This search created the first cities and led to the existence of today's enormous urban centers.

Impact of Commerce

Cities have been centers of human culture since the dawn of recorded history. Different places have different characters and histories, but common to all cities is a need for people to be in close contact for an easy flow of ideas, money and goods.

Even the language people speak is greatly influenced by economic history. In eastern Africa, Swahili is a widely spoken language that traces its origins to the interactions between Bantu-speaking people from central Africa and Arab traders on the Indian Ocean coast. English owes its status as a global language largely to the expansion of the British colonial empire in the period between 1600 and 1800.

Many early trade routes operated on land. The silk road was a famous network of routes across Asia that allowed trade between Asia and Europe. Cities across northern Africa were centers of the cross-Saharan trade. By the 1400s maritime technology was sufficiently advanced that ships could reliably cross the oceans of the world.

The colonial empires built by European powers in the period 1500–1800 created major cities across the world and set the stage for today's global economy. Many of the builders of these empires were motivated by cultural and religious

San Francisco

This high-resolution image from the IKONOS satellite shows the skyscrapers of downtown San Francisco, California. At the right are the waters of San Francisco Bay, one of the world's great harbors. Spanish explorers reached San Francisco Bay in the mid-1700s and Russian explorers approached from Alaska. A permanent Spanish presence began in 1776. When the United States took control of the area during war with Mexico in 1846, the population of San Francisco was about 600. Gold was discovered in 1849 on the American River, about 100 miles (160 km) inland from San Francisco. In a short time at least 50,000 people poured into San Francisco by land and sea. The city, with its ideal port, became the transportation hub of northern California. In 1859 silver was discovered in Nevada, and by 1870 well over 100,000 people lived in San Francisco.

reasons, but the desire to gain economic power was the most important factor. Nations competed against one another for access to the most valuable resources — wood, metals and agricultural products — which would be shipped back to the mother country. The earliest of these global empires were created by Portugal and Spain. In the early 1500s, Portuguese trading posts were built along the coasts of Africa, South America, India, China and Indonesia. Spain dismantled the Aztec and Incan civilizations and created a new empire spanning an entire continent. In the 1600s the Dutch gained control of much of the commerce formerly dominated by Portugal, and France began building an empire that would eventually have colonies on every continent. In the 1700s the British were dominant in global trade, and by the mid-1800s the British Empire was the most extensive in the world.

Shipping Across the Ocean

Ships carry the vast majority of goods traded over long distances. Sailing ships dominated the waves until steamships were developed. The first ocean-crossing steamship journey was actually a commercial failure. The *Savannah*, built in New York in 1818, was a sailing ship with an auxiliary steam engine. The ship's new owners decided to sell it in Europe. The *Savannah* stopped in Ireland, Liverpool, Stockholm and St. Petersburg, but a buyer was never found. It returned, this time using only sail power. Later attempts at transatlantic commerce were more promising. By 1855 American and British companies were running ships across the Atlantic.

Most modern shipping relies on container technology, which allows goods to be quickly transferred from sea to land transportation. About 100 million containers are carried on the seas every year. The facilities built to handle these containers are easy to identify in satellite images.

Land and Air Transportation

Canals are built to allow ships to carry products to commercial centers. The Roman Empire built extensive networks in Europe. Canals hundreds of miles long were constructed in China during the period between 300 BC and 100 AD. Modern canals, such as the Suez and Panama canals, can be seen from space as they cut through the landscape to connect seas and oceans.

Today most overland bulk cargo is carried by railroads. Railroads began service in Britain and the United States during the 1830s. By 1917 there were a million miles of railroad tracks throughout the world — some 250,000 of them

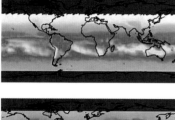

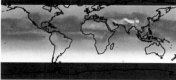

Exposure to Dangerous Rays

These maps show the amount of ultraviolet radiation reaching the Earth's surface. Sunlight includes ultraviolet (UV) rays, which can be harmful for life on Earth. Ozone in the upper atmosphere prevents most UV from reaching the Earth's surface. In these maps, yellows indicate high levels of UV exposure and reds indicate moderate levels. The amount of UV radiation reaching the surface varies with sun angles as the seasons change. These global maps can be produced almost every month. These two maps were made six months apart.

in the United States alone. The development of almost all of today's large cities was facilitated by these rail lines. Rail yards, cargo terminals and passenger stations were built in most cities and remain easily visible from space.

Highways came next in the evolution of transportation technology. Germany and Italy built the first high-speed roads in the 1920s. In 1956 the United States began building the interstate highway system, the largest highway network in the world. Eventually it included more than 44,000 miles (74,000 km) of limited-access highways. Highways and paved roads cover more ground than any other transportation system, and every large city is served by an extensive network of highways.

Air travel is the most recent transportation technology to leave its mark. Airports and support facilities cover huge amounts of land around most cities. Major airports can be larger than some cities, with thousands of employees involved in their operations. A modern airport is a city's gateway to the world, and the vast majority of long-distance travelers arrive by air. Most airports generate significant economic growth, and many cities develop airports in the hope of attracting new growth nearby.

Urban Centers

Cities have always been centers of cultural life. They are the places where people of different views and backgrounds come together from wide areas. Most cities have been built near routes on rivers or the ocean, or along extended overland routes. About 40 percent of the Earth's large cities are located on an ocean.

Because cities develop at the intersection of transportation routes, centers located where water and land transportation routes come together will generally thrive. Overland transportation routes converge at Chicago because they are required to divert around Lake Michigan. Many features form barriers to movement, like high mountains or bodies of water. If it were easy to travel on the Earth's surface in any direction, there would not be cities as we know them.

As late as the 1700s the vast majority of people lived in rural areas. There were very few cities with a population greater than 100,000. Then, in the late 1700s, the industrial revolution began sweeping across northwestern Europe and soon reached North America and Japan. Cities grew quickly as industry attracted people for employment. The building of huge factories necessitated the creation of a network of support facilities, including roads and rails to transport raw materials and finished goods.

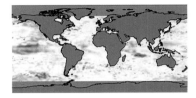

Sea Surface Height
The level of water in the Earth's oceans does not remain constant. It changes in response to tides, ocean currents and temperatures. Long-term changes may also be caused by melting polar ice due to rising global temperatures. Because the vast majority of commerce occurs on the shores where water and land meet, an understanding of sea level dynamics is very important. This map shows sea level deviations from the expected values. Blue indicates sea levels lower than normal, yellow higher than normal. The radar sensor on the TOPEX/Poseidon satellite made these measurements.

Understanding Cities

Geographers examine cities by producing models that describe the complex layout of urban areas. The most intense urban development occurs in the center of a city, where transportation routes are most accessible and the costs of new development were lowest when the city was founded. This area, the city's downtown, is known to urban geographers as the central business district. Different models describe how development extends out from the center.

One model of urban structure describes concentric zones of decreasing density. As a person travels farther away from the center, the downtown area gives way to dense industrial and residential areas that, in turn, are surrounded by low-density residential suburbs and, finally, rural areas. Another way of understanding cities is by dividing the city into sectors. Urban development is concentrated around routes of transportation reaching out from the central business district.

Models of urban development became more complex as modern cities developed into metropolitan areas with central cores and several outlying centers. In the 1990s many geographers began to describe the decentralized nature of metropolitan growth, with small centers spread across vast areas between suburban regions.

Satellite Images of Commercial Centers

Many parts of the urban structure are visible in satellite images from space. In the center of large cites, commercial skyscrapers tower over surrounding areas. Bridges and other large urban structures are easy to identify. Industrial zones and residential areas are also clearly visible.

Some recent satellites have been designed specifically to observe the Earth's surface at high resolution. These commercial satellites, which can look at objects as small as individual vehicles on the street, are well suited for observing cities and individual structures. With the variety of satellites now in orbit, it is possible to observe entire metropolitan areas or city blocks.

Planning the Future

The future of cities is the focus of the work of urban planners. Planners help to decide which areas should be designated for future growth and which should be protected from new development. Every city and local political jurisdiction has some sort of planning process. The purpose of such planning is to guide future growth. Some cities have strict laws about what can be built and where, but others have only general rules. Zoning laws may limit the density of housing that can be

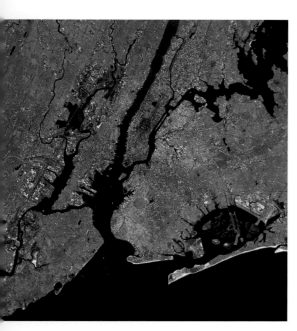

New York City

New York is shown in this image from the Landsat 7 satellite. The city began as the Dutch trading post of New Amsterdam in 1629. It grew quickly as ships arrived to trade goods from all around the Atlantic. A decade after its founding, a Dutch official reported that a quarter of the buildings in New Amsterdam were taverns catering to sailors. The city's commercial character continued after the British took the city in 1644 and renamed it New York. The city boomed after the American Revolution, and by the late 1800s more goods were flowing through New York than all other U.S. ports combined. As New York grew to become the world's largest city, bridges and tunnels were built connecting Manhattan to Brooklyn, Queens, the Bronx and New Jersey.

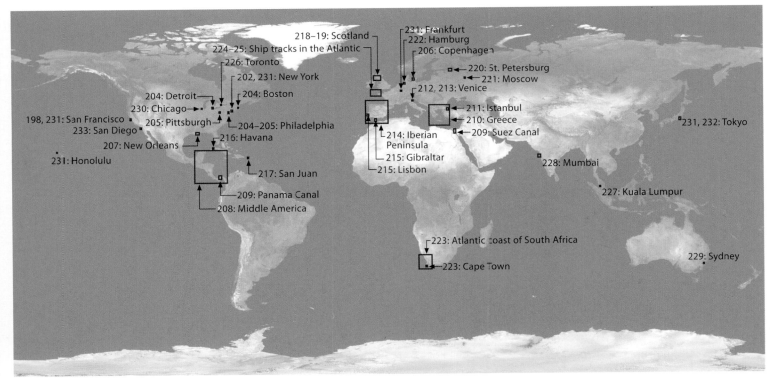

Numbers indicate page locations of the images in this chapter.

built in certain areas. Planning rules also usually separate large industrial zones from residential areas. Other zoning laws are put in place to protect open spaces or parklands, or to encourage development in certain areas.

To help understand how the city is laid out, a planning department maintains image databases of its jurisdiction. Maps created from images are used to help planners make land-use and other decisions. Until recently photographs taken from aircraft were generally used for such purposes. The new breed of high-resolution satellite images, however, can now be used in their place.

The Changing City

Central cities are still the dominant centers of commerce and population. Even as the populations of central cities have stabilized or declined, they have remained the economic and residential capitals of their metropolitan areas.

Today's cities are much bigger than ancient ones, but they obey the same rules. Cities built near transportation routes will prosper; those built in remote areas will not. Long ago, trade goods rarely traveled very far from their home ports. In today's global economic system goods, services and money are exchanged over vast distances. The growth of this dynamic global economic system is driving the next wave of changes on the Earth's cities.

Ocean Ports and River Cities

DETROIT (ABOVE)

This Landsat 7 image from December 11, 2001, shows Detroit, Michigan, with Lake Erie to the south and Lake St. Clair to the east. Colors in the lakes and river indicate sediments suspended in the water. Its location at the intersection of land and water trade routes made Detroit a prosperous city. For centuries, the Ottawa people used the Detroit River, connecting Lake Erie and Lake Huron, as a trading route. In 1701 the French built a fort and fur-trading post at Detroit, which grew to become the largest French settlement on the Great Lakes. The British captured the fort in 1760 and surrendered it to the United States at the end of the Revolutionary War in 1796. In the early 20th century, the city established itself as the center of the U.S. automobile industry.

BOSTON (TOP)

Boston was founded in 1630 by Puritan settlers from England. By 1700 it was the third-busiest port in the British Empire, behind only London and Bristol. Events in Boston were key to the American Revolution. In 1773 protesters boarded ships and dumped the cargo of English tea into the harbor. The site of the Boston Tea Party, as the protest came to be called, is at the intersection near the upper-left corner of this satellite image.

As Boston grew, it completely filled the Shawmut Peninsula. In the 1800s hills around town were removed and the earth was used to create new land. The original coastline was located where the elevated highway lies. A more recent change in Boston's landscape is the Big Dig, which replaced the highway with a tunnel. Construction was ongoing when the QuickBird satellite collected this image on December 27, 2001.

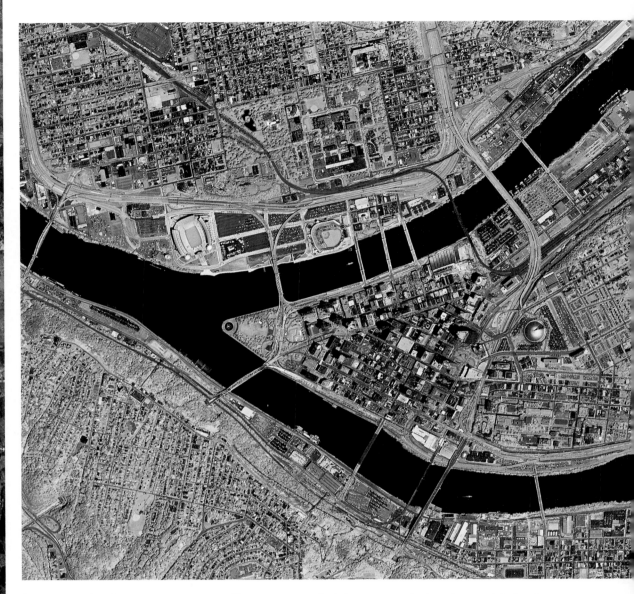

PHILADELPHIA (LEFT)

The Delaware River flows through this image. The river valley was home to Lenni Lenape people for thousands of years. In the 1600s Swedish colonists built several forts along the river. Control of these forts passed first to the Dutch and then to the British. Philadelphia was laid out between the Delaware and Schuylkill rivers in 1682, making it the first planned city in the United States. In less than a century Philadelphia had a population of 30,000 and was the largest city and most important commercial center in North America. For ten years, between 1790 and 1800, Philadelphia was the capital of the United States. This image was acquired by the ALI sensor on the EO-1 satellite, an experimental mission to test new technologies

PITTSBURGH (ABOVE)

This image was collected by the EROS A satellite on June 23, 2002. The city of Pittsburgh is in western Pennsylvania, where the Allegheny and Monongahela rivers come together to form the Ohio River. The confluence of rivers made it a strategic site for the European powers vying for dominance of North America in the 18th century. The French built Fort Duquesne on the site in 1754. Captured by the British just four years later, it was renamed Fort Pitt and began to attract settlers. The town of Pittsburgh was laid out in 1764. The site of Fort Pitt is visible in the satellite image at the point where the three rivers meet. In the 1800s Pittsburgh became one of the most important industrial centers in the United States. The rivers still play an important role in the city's economy. Despite being far from the ocean, Pittsburgh is one of busiest ports in the United States. It handles more domestic freight than any other port except for New York, Houston and greater New Orleans.

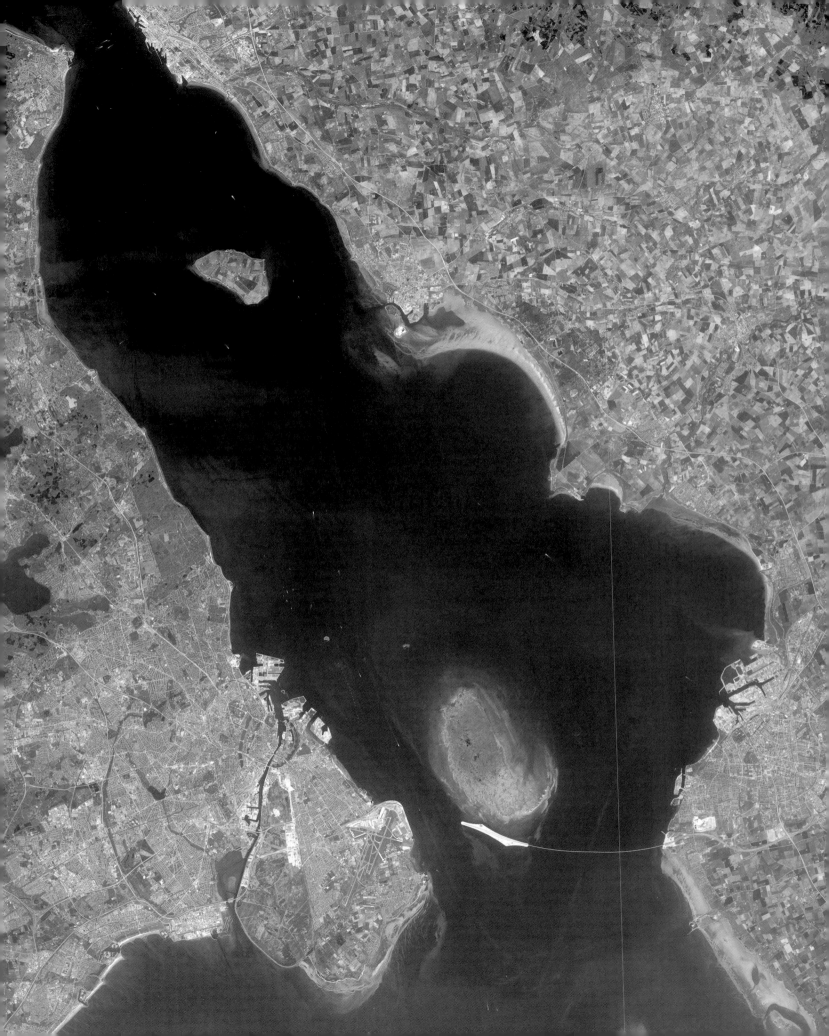

Øresund Bridge

COPENHAGEN (LEFT)

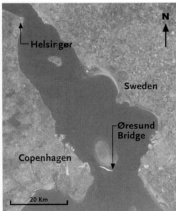

This image from the SPOT 5 satellite shows the Øresund Bridge, which connects Copenhagen, Denmark, and Malmö, Sweden. The bridge carries road and rail traffic above the water. On the west side, traffic drops into a tunnel dug underneath the water. The bridge allows people and goods to move freely across the strait between the two European Union members.

Denmark was the home of the Vikings, who expanded across Europe in the years between 800 and 1100 AD. Viking settlements and trade routes extended eastward across Russia and westward across the Atlantic to Newfoundland. At the center of Copenhagen is a channel giving ships access to the North Atlantic and the Baltic Sea.

The city of Helsingør lies on the Danish side of the channel north of Copenhagen. Also known as Elsinore, it is the setting for Shakespeare's *Hamlet*. In the 1990s ferries that cross the strait at Helsingør became popular with Swedes wanting to take advantage of Denmark's lower taxes on beer and liquor.

Lake Pontchartrain Causeway

NEW ORLEANS (ABOVE

To the north of New Orleans stretches the Lake Pontchartrain Causeway, connecting the city to suburbs on the northern shore. At 24 miles (39 km), it is the longest bridge in the world.

The causeway consists of two parallel roadways, the first completed in 1956 and the second in 1969. It is a popular resting spot for migratory birds. This image was collected by the Landsat 7 satellite on April 26, 2000.

The Mississippi River is shown here flowing past New Orleans. The city became the fourth-busiest port in the world in the 1840s, and ports in southern Louisiana still handle more freight than any other area in the United States. Surrounded by water on all sides, New Orleans is particularly susceptible to flooding. Serious damage in the aftermath of Hurricane Katrina in 2005 can be seen on page 261.

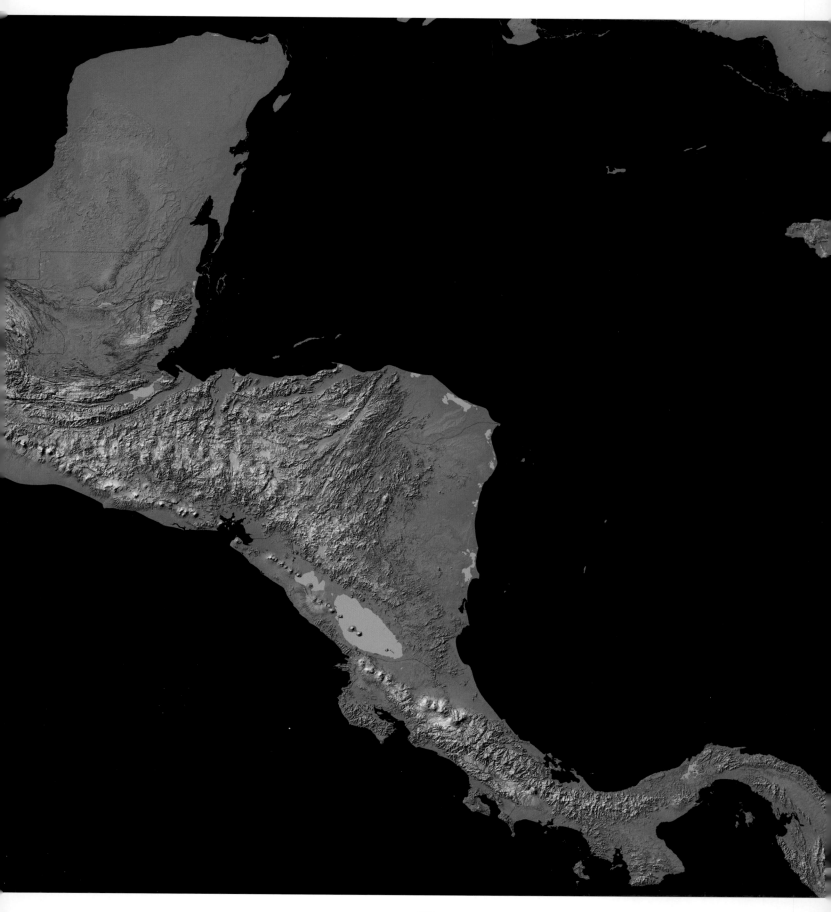

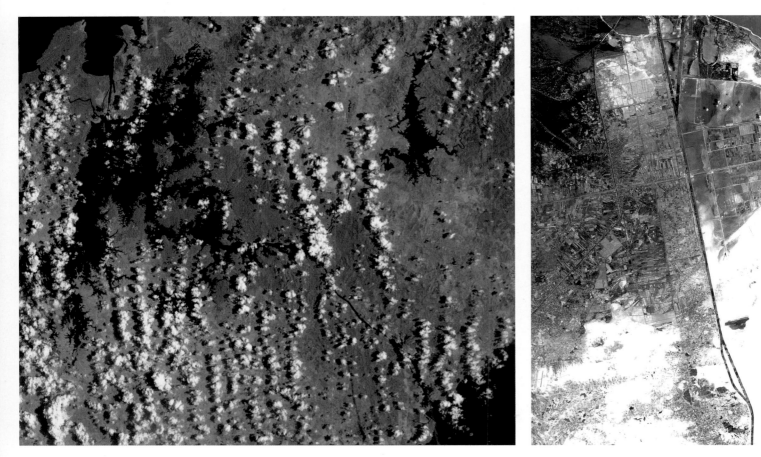

Canals

MIDDLE AMERICA (LEFT)

PANAMA CANAL (ABOVE LEFT)

The long isthmus connecting North and South America is visible in the topographic view on the left produced using orbital radar data from the Space Shuttle Radar Topography Mission. Green indicates low elevations, white high elevations, and brown the areas in between. Shading was added to accentuate vertical relief.

It was long recognized that a canal across the isthmus would benefit commerce. Two possible routes had been identified: one through Nicaragua, and one through Panama. A Nicaraguan canal would have followed the course of the San Juan River, then gone through Lake Nicaragua and into the Pacific. A French company began construction of a canal through Panama in 1881. Tropical diseases wreaked havoc among the workers, and the effort was halted within eight years as financial losses mounted. In 1903, with American assistance, Panama declared independence from Colombia. Three months later Panama signed a treaty leasing a swath of territory to the United States. Up to 40,000 people worked to complete the Panama Canal in 1914.

The image above was collected by the Landsat 4 satellite on April 1, 1990. It shows the Panama Canal as seen through the clouds. In the center lies Gatun Lake,

which was the largest artificial lake on Earth when the canal was built. Within the lake is Barro Colorado Island, a site used by the Smithsonian Institution for tropical forest research. Locks carry ships up and over the center of Panama, as the ships are pulled along by track-mounted engines. When the canal first opened, about 800 ships used it annually. By 2000 the annual total was over 10,000. It takes only about 15 hours for a ship to pass through the canal and saves a journey all the way around South America.

SUEZ CANAL (ABOVE RIGHT)

In Egypt, the Suez Canal was cut between the Mediterranean and the Red Sea in a decade-long construction project that began in 1859. The northern part of the Suez Canal is visible in this image acquired by the ASTER sensor on the Terra satellite, May 19, 2000. Extensive salt marshes are near the Mediterranean

end of the canal to the north. Dark red and blue indicate agricultural fields. The light areas are bare sand.

The Suez Canal, one of the most important waterways in the world, connects Europe with eastern Africa and southern Asia. It is 121 miles (195 km) long and at least 197 feet (60 m) wide. When it first opened the canal was less than half that width. More than 15,000 ships make the trip through the canal every year. Unlike Panama the land here is flat, so no locks are required. Canals have been built in this area since ancient times. A lake in the middle of the present canal was connected to the Nile around 1850 BC. The Greeks and Romans built canals that reached to the Red Sea, allowing trade to flow to the south.

Early Centers of Global Trade

GREECE (LEFT)

Spread out over more than 2,000 islands and a mainland of long coastlines, Greece is a nation built around the sea. Greece and parts of Turkey, Italy, Albania and Macedonia are visible in this satellite image, collected by the MODIS sensor on the Terra satellite on June 6, 2001.

Boats were sailing the Aegean Sea some 14,000 years ago. The Minoan civilization on the island of Crete developed extensive trade networks around 3000 BC. By 600 BC Greek trading settlements surrounded the Mediterranean and Black seas. In 335 BC the new king of Macedonia, Alexander the Great, began his famous military campaign against the Persian Empire, centered in modern Iran. Alexander's generals controlled territory all the way to India, and, through their trade networks, they left a lasting impact on European and Asian cultures.

ISTANBUL (ABOVE)

Istanbul occupies both sides of the Bosporus Strait, which connects the Black Sea to the Mediterranean. This is traditionally a meeting place between Europe and Asia. For centuries, trade routes reaching halfway across the world converged here, making it one of the richest cities in the world.

Istanbul was established around 650 BC as a Greek trading colony called Byzantium. It became the largest Greek trading center and eventually grew larger than any city in Greece itself. Byzantium came under Roman rule in 196 AD. In 330 it was renamed Constantinople and became the new capital of the Roman Empire. Later it was the capital of the Byzantine Empire.

The city was conquered by the Ottoman Empire in 1453. In later times the city became known as Istanbul in Arabic and Turkish. Today Istanbul is the largest city in Turkey. This image was acquired by the ASTER sensor on the Terra satellite on June 16, 2000.

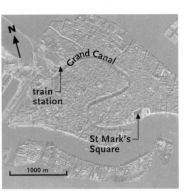

Early Centers
of Global Trade

VENICE (LEFT)

VENICE LAGOON (ABOVE)

The famous canals that cut across Venice, Italy, are visible in the image on the left, taken on November 27, 2000, from the IKONOS satellite. Running across the center of the city is the Grand Canal, winding for 2 miles (3.2 km) from St. Mark's Square to the train station in the northwest. Hundreds of small canals branch off the Grand Canal to the east and west. There are 114 canals in Venice and more than 400 bridges to cross over them.

First established about 600 AD on a small island at the edge of the Byzantine Empire, the city prospered as a commercial connection between mainland Europe and the Mediterranean. In 1082 Venice was allowed to conduct free trade throughout the Byzantine Empire, the only city with that power. During the next century the center of Europe's trade with Asia shifted from Constantinople to Venice.

Venice was built on a marshy lagoon between barrier islands and the mainland. The city is slowly sinking, and buildings are suffering the effects of water seeping into foundations. The image above, acquired by the India Remote Sensing satellite, shows water flowing out of the lagoon and into the ocean during a low tide. A long plume of sediment is visible moving through a gap in the barrier islands called the Port of Lido.

A system of floodgates, under construction across this gap, should protect Venice from flooding during high tides. It may also raise concentrations of sediments and pollution by preventing circulation of water between the ocean and lagoon.

Centers of Sea Power

IBERIAN PENINSULA (LEFT)

This image, collected by the MERIS sensor on the Envisat satellite on March 23, 2002, shows the Iberian Peninsula. Yellow indicates arid areas at higher elevation. White cloud cover dominates the upper part of the image.

The Iberian Peninsula is home to Spain and Portugal, the world's first global sea powers. Their location gave these nations a natural advantage when Europeans began to cross the world's oceans in the early 1500s. Centuries before, the Iberian Peninsula was also the place where African and European civilizations met. In 711 AD armies poured across Gibraltar into what is now Spain, taking control of most of the peninsula and briefly crossing the Pyrenees into southern France. For the next 700 years most of the Iberian Peninsula was under the control of a series of Islamic states and contained the most dynamic and cosmopolitan cities in Europe.

LISBON (ABOVE LEFT)

This image from the IKONOS satellite shows the historic port area of Lisbon, the capital and largest city of Portugal. Ships leaving from Lisbon explored the coast of Africa in the early 1400s. Portugal created the world's first large-scale sugarcane plantations in the Azores and Madeira Islands in the 1420s. During the 1500s a series of Portuguese trading posts and colonies — the most extensive trading network the world had ever seen — was established along the coasts of Africa, the Middle East, South America, India, China and Indonesia.

GIBRALTAR (ABOVE RIGHT)

The Strait of Gibraltar separates Europe from Africa at the western end of the Mediterranean Sea. This image of the strait was collected by the ASTER sensor on the Terra satellite on July 5, 2000. Spain lies to the north, Morocco to the south. On the north side of the strait lies the Rock of Gibraltar. Rising 1,398 feet (426 m) above the sea, it sticks out on the thin strip of land on the north side of the strait. As the connection between the Mediterranean Sea and the Atlantic Ocean, the Strait of Gibraltar, only about 27 miles (43 km) wide, has always been an important military and commercial pathway for shipping. In 1704 the British took control of the Rock. Gibraltar remains British and Spain holds small enclaves on the Moroccan side of the strait.

New World Ports

HAVANA, CUBA (LEFT)

Havana, with more than two million people, is the capital and largest city in Cuba. Old Havana, with buildings dating to the 1500s, lies near the center of this image. The harbor at Havana is the largest in the Caribbean. At least eight ships are visible in this image captured by the IKONOS satellite.

The Spanish founded the city in 1519. In 1553 it became the capital of Cuba and was one of the most important ports in Spanish America. In 1634 the king of Spain issued a decree calling Havana the "Key to the New World and Rampart of the West Indies." Indeed it was. Havana was the first port of call for ships from Spain and the starting point for major expeditions to Mexico and South America. As Spanish forces conquered the Aztec and Inca empires, the wealth of an entire continent flowed through Havana and made Spain the richest nation in the world.

SAN JUAN, PUERTO RICO (ABOVE)

San Juan is the largest city on the Caribbean island of Puerto Rico and an important Caribbean transportation hub. This image from the IKONOS satellite shows cruise ships docked in the harbor and Fernando Luis Ribas Dominicci Airport in the lower right.

The Spanish founded San Juan in 1508. It is home to San José Church, the oldest Christian religious structure in the Western Hemisphere. Many groups sought control of San Juan. The Carib people attacked first, followed by the Dutch and British. The fortresses La Fortaleza and El Morro were built in the early 1500s. In the 1630s construction began on San Cristobal, the largest Spanish fort ever built in the Americas. All three forts still stand. Puerto Rico became part of the United States in 1898 after war with Spain.

Sea and Land

SCOTLAND

This color infrared image, acquired by the Landsat 7 satellite, shows central Scotland. Urban areas are blue, forests are dark red, and grasses and agricultural fields are orange. Areas with little vegetation at high elevations are gray.

Both Glasgow and Edinburgh developed at the intersection of sea and land transportation routes. The Scottish highlands are to the north, and the Scottish lowlands are to the south. The Clyde River reaches to the west, and the Firth of Forth to the east. Glasgow is Scotland's commercial center and largest city. Edinburgh has long been the center of government. Trade with America fueled Glasgow's growth in the 1700s, first in tobacco and later in cotton. The Clyde was dredged to provide access to the sea, and Glasgow soon became a major shipbuilding center.

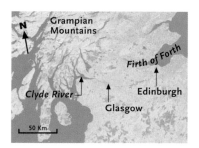

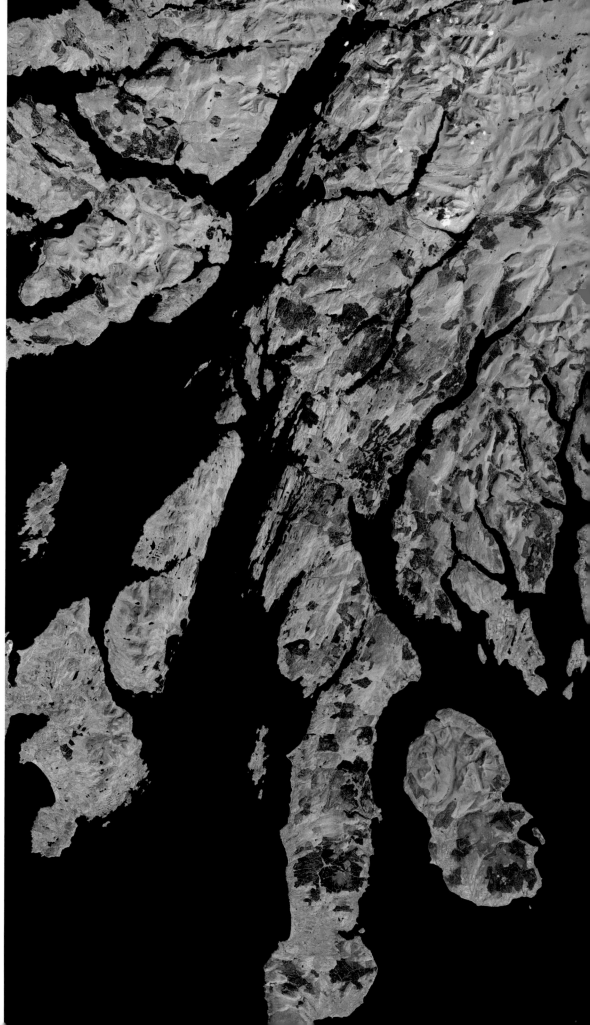

Trading along Rivers

ST. PETERSBURG (LEFT)

This image was collected by the SPOT 4 satellite on August 26, 1996. St. Petersburg, the second-largest city in Russia, lies to the west. The Neva River connects the Gulf of Finland to Lake Ladoga.

The city was founded in 1703 by Peter the Great, the czar of Russia, as a new capital city and commercial port to connect Russia to the dynamic European economy. The site chosen was a natural gateway. Within six years a series of canals connected St. Petersburg with rivers draining into the Volga, and by 1726 more than 90 percent of Russian foreign trade flowed through St. Petersburg. Industries grew up around new naval facilities. By the time it was a century old, St. Petersburg had a population of about 250,000. The Russian Revolution began in St. Petersburg during World War I, leading to the creation of the Soviet Union. The capital was later moved back to Moscow.

MOSCOW (ABOVE)

Moscow is the capital and largest city of Russia. This image was acquired by the EROS A satellite on March 18, 2002. The Moscow River flows into the Volga, allowing goods to travel all the way to the Caspian Sea. Roads are arranged in concentric circles that mark the location of fortifications built in the late 1500s. Visible in the image are the Boulevard Ring and the Garden Ring.

At the center of town is the famous Kremlin. This fortress was built in the 12th century where the Neglinnaya River joined the Moscow River. The Neglinnaya has since been paved over and flows in pipes beneath the streets. A large trading settlement developed east of the Kremlin in the 1300s, and Moscow grew to become the largest city in Russia.

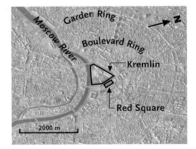

On the east side of the Kremlin is Red Square, one of the largest public spaces in the world. As the commercial heart of the city, Red Square hosted Russia's largest public market, with goods from western Europe and central Asia passing through its stalls. The square, known as Red or "Beautiful" Square since the 1600s, has also been a center of Russian cultural life for centuries.

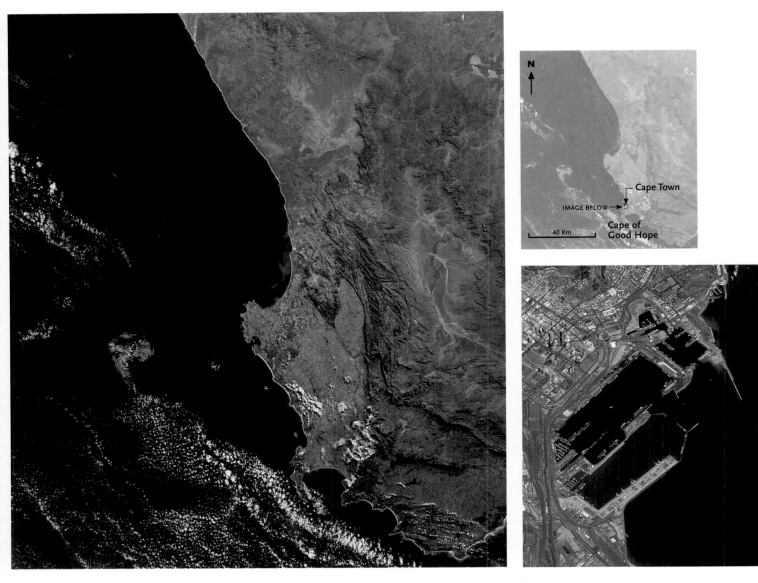

N

Cape Town

IMAGE BELOW

40 Km

Cape of
Good Hope

Modern Shipping Technology

HAMBURG, GERMANY (LEFT)

Modern port facilities use container shipping to move goods from ships to trains and trucks. This image of Hamburg, Germany, shows the largest container-shipping terminal in Europe. Cranes remove containers from ships and transfer them to the large paved holding areas found on both sides. In the lower left, rail lines transport containers across Europe. This image was collected by the QuickBird satellite on May 10, 2002.

Hamburg is the second-largest city in Germany after Berlin. It stands on the Elbe, where the river widens before emptying into the North Sea. Today,

people from all over the world pass through this hub of global commerce. Only New York City has a greater number of foreign consulates.

ATLANTIC COAST OF SOUTH AFRICA (ABOVE LEFT)
CAPE TOWN, SOUTH AFRICA (ABOVE RIGHT)

The above image, acquired by the MODIS sensor on the Terra satellite on May 5, 2003, shows the Atlantic coast of South Africa. Near the bottom of the image is the Cape of Good Hope. Mountain ranges separate the low-lying coastal terrain from the higher ground in the interior.

The upper-right image, collected by the IKONOS satellite, shows Cape Town,

South Africa. In 1652 Dutch ships landed at the present site of Cape Town. Construction of a fort began in 1666. This structure, the oldest standing building in South Africa, is surrounded by the modern city. When the British took control of the Cape in the early 1800s, about 15,000 people lived in the town. In 1870 diamonds were discovered in the interior of South Africa. Then gold was discovered, about 16 years later. The city grew as port facilities were expanded to handle more cargo. Cape Town remains South Africa's leading port.

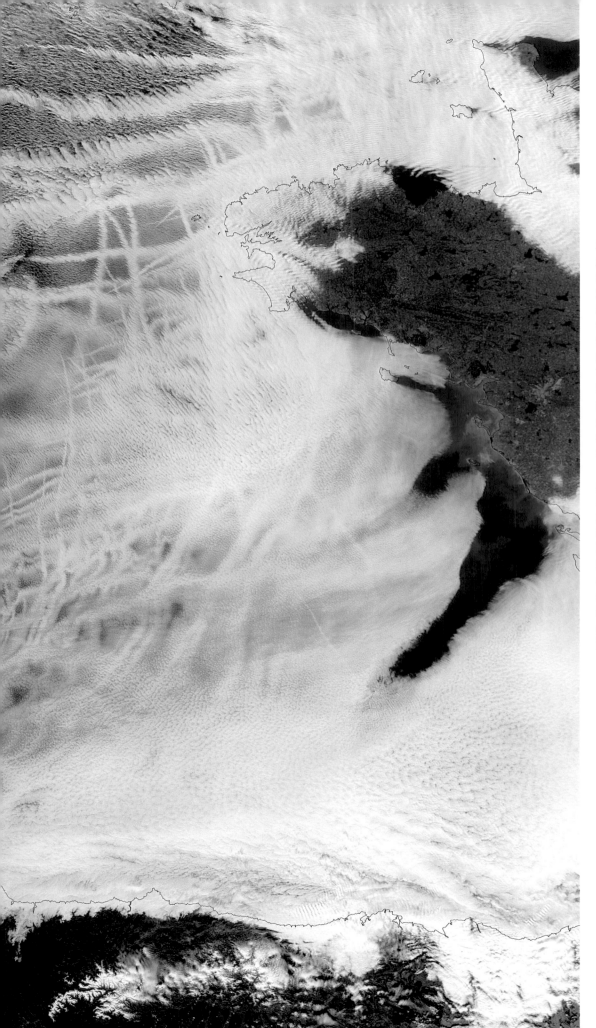

Ship Tracks

ATLANTIC OCEAN

Ships cross back and forth across the Atlantic carrying goods between North America, Europe and the rest of the world. This image shows dozens of tracks left by ships traveling the North Atlantic between France and England. The coastline has been drawn on this image. The density of tracks demonstrates the heavy volume of commerce moving on the Atlantic. This image was acquired by the MODIS sensor on the Aqua satellite on January 27, 2003.

Ship exhaust contains microscopic particles that mix with water molecules in the atmosphere. If the density of particles is high enough, small clouds form around exhaust plumes, leaving long, thin clouds in the atmosphere. Although they look like natural clouds, contrails and ship tracks are made of smaller water droplets. This causes them to reflect more sunlight, which cools surface temperatures and makes them less likely to create precipitation.

Reaching to the Sky

TORONTO (ABOVE)

Toronto is the largest city in Canada. This image of downtown Toronto shows the CN Tower, the world's second-tallest free-standing structure after the Indostat Telecom Tower in Jakarta. The tower is 1,815 feet (553 m) high and contains transmission antennas and an observation deck. The tower stands next to the SkyDome stadium and the railroad tracks leading to Union Station. Toronto Harbour is visible at the lower edge. This image was acquired by the IKONOS satellite on March 18, 2000.

In the 1600s the Seneca people, members of the Iroquois Confederacy, built a trading port on the Humber River west of downtown Toronto. French fur traders arrived in the mid-1700s. In 1763 Britain took control of Canada, and 30 years later the small settlement of York

was chosen as the capital of Ontario. In 1834 the city was renamed Toronto. The railroad arrived in 1850, and the city grew quickly to become a major port on the Great Lakes.

KUALA LUMPUR (RIGHT)

Kuala Lumpur is the capital and largest city of Malaysia. The city has become an important financial center for southeast Asia. This image from the IKONOS satellite shows the Petronas Towers, headquarters of Malaysia's national petroleum company. At 1,483 feet (451 m) they are the second-tallest office buildings in the world. The tallest is the Taipei 101 Building in Taipei, Taiwan, completed in 2004.

In the 1400s the city-state of Malacca, a busy port just south of Kuala Lumpur, was home to thousands of merchants. Chinese fleets on their way

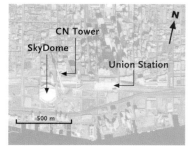

to the Indian Ocean stopped there. In 1511 a Portuguese fleet took control of Malacca. Today, nearby Singapore is a global shipping center.

Malaysia became British territory in the 1800s. Chinese and Indian immigrants flooded in during the colonial years. In 1857 a tin-mining settlement was developed at the confluence of the Kelang and Gombak rivers. Later known as Kuala Lumpur, the city grew as railroads reached the city in the 1880s.

Ocean Port Cities

MUMBAI, INDIA (LEFT)

Mumbai, with a population of over 12 million people, is the largest city on the west coast of India. The city was known as Bombay until 2002. The harbor hosts ships crossing the Arabian Sea to Africa, Europe and the rest of the world. Over a third of India's foreign trade passes through Mumbai. The peninsula upon which the city lies was made up of seven islands. Since the 1600s the islands have been connected with landfill. In this Landsat 5 image from November 9, 1992, green areas indicate vegetation cover, light brown areas are lightly covered with vegetation, and bright blue indicates urban areas.

In 1348 this area became part of the Gujarat Kingdom, and the Portuguese took control in 1534. They named the bay Bom Bahia ("beautiful bay"), which later became Bombay. In 1664 it was peacefully transferred to British control. Bombay's cotton and textile business grew during the American Civil War in the 1860s, when Britain was forced to look for new supplies of cotton. Still more trade flowed through Bombay when the Suez Canal opened in 1869.

SYDNEY, AUSTRALIA (ABOVE)

Sydney, the largest city in Australia, possesses one of the finest harbors in the world. This image shows the downtown area crowded with commercial towers. Ships can be seen in the harbor. The famous Sydney Opera House, completed in 1973, juts out into the harbor, and the Sydney Harbor Bridge can be seen crossing the water to the right. This image was acquired by the QuickBird satellite on April 4, 2002.

The first British expedition arrived in 1788 to deliver convicts to a new penal colony. It was not long, however, before Sydney became an important trading center. By the early 1900s Sydney surpassed Melbourne as Australia's largest city as trade links with China and North America became more important than links to India and Europe.

Air Travel

Today, most long distance travel is by airplane. All large cities must allow a great deal of space for runways and facilities to accommodate large aircraft and the people they carry.

The image on the opposite page, acquired by the IKONOS satellite, shows Chicago's O'Hare International Airport. It competes with Hartsfield-Jackson Atlanta International Airport in Atlanta for the title as the world's busiest airport. Several runways surround the central terminal area. There are plans to expand to the west with additional runways.

The upper left image shows San Francisco International Airport, with runways built out into San Francisco Bay. The lower left image shows the nine terminal buildings at John F. Kennedy Airport in New York, an important gateway into the United States. About 37,000 people work at JFK airport, which handles about 30 million passengers each year.

The upper right image shows Frankfurt International Airport, the busiest airport in Germany and the second busiest in Europe after London's Heathrow Airport. The image on the right from the EROS A satellite shows Honolulu International Airport, with a major runway built into the ocean. The lower right image shows Tokyo's Haneda Airport on Tokyo Bay. The image was acquired by the OrbView-3 satellite on September 27, 2003.

▲ SAN FRANCISCO INTERNATIONAL AIRPORT

▲ NEW YORK JOHN F. KENNEDY INTERNATIONAL AIRPORT

◄ CHICAGO O'HARE INTERNATIONAL AIRPORT

▲ FRANKFURT INTERNATIONAL AIRPORT

▲ HONOLULU INTERNATIONAL AIRPORT

▲ TOKYO HANEDA INTERNATIONAL AIRPORT

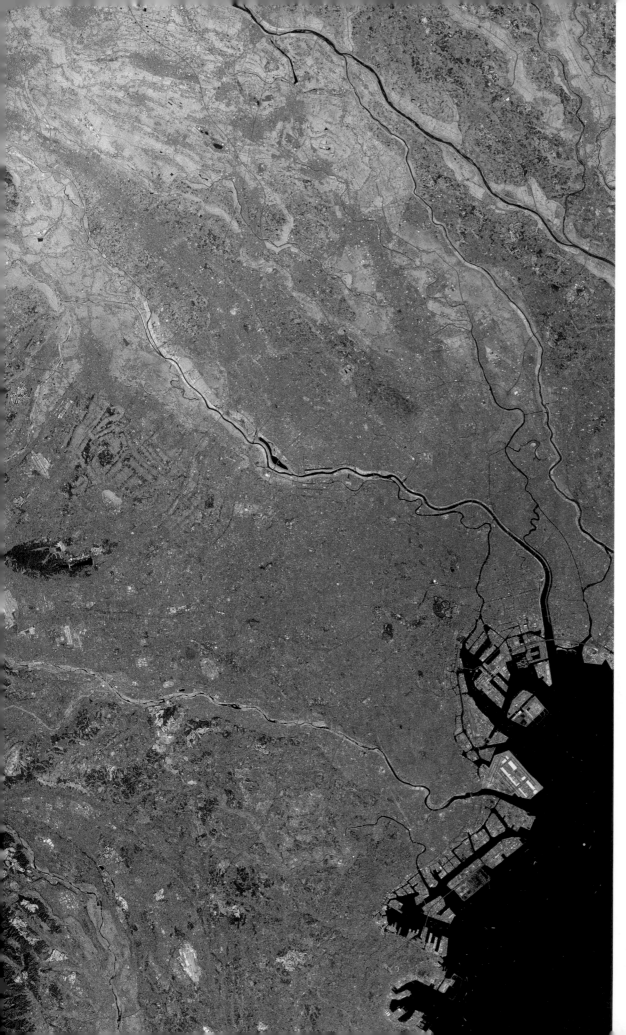

Trading Around the Pacific

TOKYO (LEFT)

Tokyo is the largest city in Japan, and Tokyo Bay, in the lower part of the image, is one of the world's largest and busiest harbors. To the south are Kawasaki and Yokohama, Japan's second-largest city. By some measures, these cites form the largest metropolitan area in the world.

Tokyo was originally called Edo, a walled city built in the 15th century near the banks of the Sumida River. During the 1700s Edo was probably the first city in the world to reach a population of over one million. The oldest part of the city is just east of the Imperial Palace. In the 1900s surrounding villages grew into major centers. One of the largest, Shinjuku, is now home to office buildings, shopping districts and the busiest train station in the world. This image was collected by the ASTER sensor on the Terra satellite on March 22, 2000.

SAN DIEGO (RIGHT)

Now a major Pacific port, San Diego began as a small Mexican town near the site of a Spanish mission. The area was annexed to the United States in 1846, and the streets of modern San Diego were first laid out in 1867. The city grew quickly after the railroad arrived from the east in 1884.

Buildings in downtown San Diego are visible in this image from the IKONOS satellite, as is a marina on the water. The green square in the upper part of the image is Balboa Park, home to many of the city's cultural institutions.

Dynamic Earth

There is one constant on the Earth: change. The atmosphere, the land surface and life forms all undergo changes through time. Vegetation on the Earth's surface changes in response to climate variations. Rivers and lakes can dry up or overflow. Winds blow sand and dust across the surface. Large changes in the composition of the atmosphere occur over long periods. Some changes are part of a cycle that may have begun millions of years ago. Other changes, including those caused by human activity, occur much more quickly.

Satellite remote sensing platforms are uniquely suited to detecting changes on a global scale. As the satellite orbits the planet, large areas can be observed and revisited many times. The orbits of most satellites are designed so that they regularly pass over the same part of the Earth's surface. This allows scientists to compare images taken at different times. Global dynamics such as atmospheric changes can be measured on the ground, but only with the satellite perspective is it possible to obtain a broad enough view to understand such changes. The long-term effect of human activity on the Earth is not well understood, but remote sensing is the most promising tool for providing answers.

Changing Seasons

Many areas on the Earth experience wide variations in temperature and moisture as the seasons change. These changes affect vegetation on the surface by allowing trees and grasses to thrive in the spring and summer and become less active in the autumn and winter. Because satellite sensors are well suited for detecting vegetation, they can produce images showing how these changes take place. Many such images can be combined to build global maps showing the onset of spring and the approach of winter.

Although it is possible to detect changes over periods as short as a few weeks, a common method of detecting change is to compare satellite images collected years apart during similar seasons. Comparing images taken during summer and winter would show only

Disappearing Forests

BOLIVIA

This Landsat 7 image (opposite page) shows part of Bolivia on August 1, 2000. The pattern of brightly colored shapes shows how farmers have divided the land into individual fields. Bright red indicates growing crops. Also visible, as dark strips, are windbreaks, which can prevent the fine soils from being lost to wind erosion. This area was once entirely covered by dense forest. Roads were built into the area to accommodate logging operations, ranching and new settlements. The only forests remaining in this image are in the upper- and lower-right corners. The colors in the image were created by infrared light reflected from vegetation cover and bare ground. The pie-shaped fields on the northern part of the image are a development project where community centers are built at the center of each circle. This area is also shown in a sequence of images on page 237.

seasonal changes, and not land dynamics. Scientists compare summer images from different years if they are interested in seeing vegetation changes. Images collected in the winter are sometimes compared to detect surface features not related to vegetation. These are often called "leaf-off" images, because deciduous trees lose their leaves during this time of year.

Regional Changes

Large bodies of water experience some of the most drastic changes visible on the Earth's surface. Rivers often spill over their banks, flooding vast areas. Some large lakes are facing dropping water levels. Many satellite sensors are able to detect water on the surface, and the data can then be used to produce maps of the changes.

In arid areas dust can be kicked up by strong winds. Dust storms are common when atmospheric patterns change over large deserts. During these storms, visibility disappears and all activity must wait for the dust storm to pass.

Fire is a common agent of change on Earth. Fires are a natural part of the forest life cycle, removing older growth and allowing room for new growth. When large fires break out they quickly become visible to remote sensing instruments, which can detect the thermal-infrared emissions from burning forest fires. Satellite images can also be used to track the plumes of smoke.

Landcover Change

The Earth's surface is covered by many different types of material. Some are natural, such as trees, grass, rocks and ice. Some are produced by humans, such as those used in highways, houses and industrial areas. The categories of materials are known as landcover types. Knowledge of landcover types is important for understanding how the Earth's surface absorbs solar radiation and interacts with the atmosphere.

Landcover types can have an important effect on environmental processes. Forested areas absorb sunlight and carbon dioxide from the atmosphere, while trees circulate water between the soil and air. In contrast, a paved urban area reflects a great deal of sunlight and does not allow precipitation to enter the ground.

Because of the dynamic nature of the Earth's surface, the landcover changes over time. Many changes are part of natural cycles, but abrupt landcover changes are often driven by human activities. Cities grow and cover more land, and humans move into forested areas to clear them for agriculture.

In the United States and Canada, the most common type of landcover

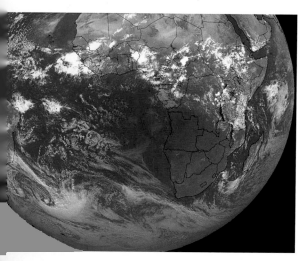

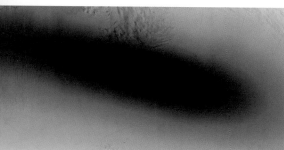

Solar Eclipses

Solar eclipses are among the most spectacular events that can be recorded from satellites. A weather satellite in geostationary orbit captured the upper image of a solar eclipse on June 21, 2001, as the Moon came between the Sun and the Earth, producing an eclipse visible in southern Africa. The lower image was acquired by the MODIS sensor on the Terra satellite on November 23, 2003, as the Moon cast a shadow about 300 miles (500 km) long across the frozen surface of Antarctica.

change during the 1900s resulted from the extension of urban landscapes. In South America, expanded soybean cultivation is one of the factors driving landcover change.

One of the potentially most devastating effects of landcover change is the loss of wildlife habitat. Hundreds or perhaps thousands of species become extinct every year as a result of human-induced loss of habitat.

Using remote sensing it is possible to map different types of landcover on the Earth's surface. Sensors can discern different types of vegetation, water cover and human-made surfaces. Each type of landcover reflects a different combination of wavelengths of light back into space. This unique combination is called a spectral signature. For instance, a forest of trees reflects very strongly in near-infrared light but does not reflect as strongly in visible light. The opposite is true for most human-made surfaces. To detect these signatures, the multispectral sensors in orbit collect data from several wavelengths of light.

Landcover maps at the global scale allow scientists working in a wide variety of fields to improve their understanding of how the Earth's surface interacts with the atmosphere and global climate.

Global Cycles

A period of cool weather during the 1600s and 1700s was known as the Little Ice Age. Since then the Earth's climate has generally been warming. Scientists are still trying to understand the complex interactions that govern climate change and the effect humans are having on the process.

Although weather records have been kept for only a few centuries, scientists can use ice and soil samples containing pollen and insects to learn about the temperature and climate of thousands of years past. With these techniques it is also possible to determine the amount of oxygen and carbon dioxide that was once found in the atmosphere. Both historical records and scientific data show fluctuations in temperature and atmospheric gases over time.

There are several mechanisms that remove carbon dioxide from the atmosphere, including absorption by the oceans. Another important factor is the growth of vegetation. As forests grow, they remove carbon dioxide from the air. Many different methods are used to estimate the growth of vegetation through time, a process called primary production. Satellite remote sensing is the primary tool being used to help us understand this process on a global scale.

Tracking Deforestation
Satellite images can be used to track deforestation. This sequence of Landsat images was collected over a period of 25 years: the first is from 1975, the second from 1992, and the third from 2000. Red indicates dense forest. The lighter colors indicate cleared land, mostly soybean fields.

Santa Cruz, Bolivia's second-largest city, is in the lower left. The first paved road did not reach the city until 1950, but by 2000 Santa Cruz had a population over 900,000. Large-scale soybean cultivation began in the early 1970s, and within 20 years the crop was among the most important in Bolivia.

Changing Atmosphere

Not only does the atmosphere provide air to breathe; it also moderates surface temperature. Several gases in the atmosphere, such as carbon dioxide, water vapor and methane, are called "greenhouse gases" because they trap warmth inside the atmosphere. They allow sunlight to shine on the Earth's surface, but they do not allow heat to escape into space. Without the effect of greenhouse gases, nighttime temperatures would be about 70°F (40°C) cooler, perhaps too low to permit life to exist.

Too much of a greenhouse effect may, however, be a bad thing. Monitoring of the atmosphere has shown that levels of greenhouse gases are rising. Although carbon dioxide is only a small part of the atmosphere, it is the most common greenhouse gas. Most carbon dioxide is released by plants or produced by the decay of dead vegetation, but volcanic eruptions are also an important source. The burning of oil, coal and other fossil fuels also releases carbon dioxide into the atmosphere. Since the industrial revolution, the amount of carbon dioxide in the atmosphere increased by about 30 percent, and the amount of methane more than doubled. Today, the Earth's atmosphere probably contains more carbon dioxide than it has for millions of years, which may have important effects on global climate.

A Warming Planet

The Earth is getting warmer. The average temperature is increasing by a rate of between 1 and 3°F (0.6 and 1.5°C) per century, compared with an increase of 1.8°F (1°C) every 4,000 years since the last ice age. The warming matches the rise in greenhouse gases, and by adding carbon dioxide and methane to the atmosphere humans are playing a role in these changes.

A significant rise in global temperatures would have many effects. Warmer oceans would rise as a result of thermal expansion, and melting ice would also add water to the oceans. Estimates vary greatly about the relationship between warming and sea level, but temperatures around the Antarctic Peninsula have risen by 3.5°F (2°C) since the 1960s. A warmer planet would also see significant changes in climate. Storms would be more common in tropical areas, and prolonged droughts would be more common at higher latitudes. Warming would likely be most noticeable in northern areas of North America and Asia. Ecosystems and habitats would change, since warmer and dryer conditions would exist closer to the Earth's North and South poles.

Fires

Fires are an important agent of change in the natural world. The blue band across the map shows the area sampled, and the location of large fires appears in red. Fires are used to clear forests, making way for agricultural and other purposes. Besides altering landcover, fires affect climate and air quality.

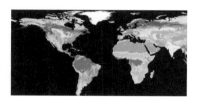

Global Landcover

This global map of landcover is derived from several sources of satellite images. The colors in the map indicate various types of vegetation or environment, including ice, desert, forest and grassland, each of which reflects light differently. At the global scale, maps like these can show environmental changes across the planet.

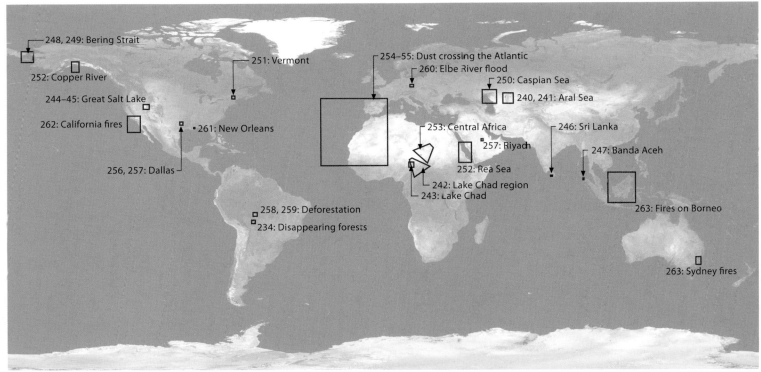

Numbers indicate page locations of the images in this chapter.

As concern about global warming increased, international agreements were proposed to slow the release of greenhouse gases in the atmosphere. These agreements have led to greater cooperation in environmental research and encouraged the use of satellite remote sensing to monitor climate changes.

The Future of Planet Earth

Atmospheric and surface changes have been modeled into the future, creating predictions of trends. Satellite imagery can be incorporated into these models to give them a better chance of accurately predicting what changes will take place.

One thing is certain: the Earth will continue to change in the future. Scientists are following global climate changes using remote sensing data on the Earth's plant life, atmosphere and oceans to help understand how global processes operate.

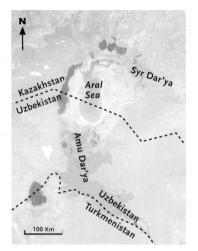

Disappearing Sea

ARAL SEA, 1998 (LEFT)

2002 (RIGHT)

The Aral Sea in central Asia was once the world's fourth-largest inland body of water. The image on the left was acquired in 1998 by the Resurs-O1 3 satellite. The image on the right, from the MODIS sensor on the Terra satellite, was acquired only four years later. By 2020, the Aral Sea will probably be only about 10 percent of its original size. The disappearance of the Aral Sea may be the most significant regional environmental disaster the Earth is facing, but no plans exist for solving the problem.

The Aral Sea straddles the border between Uzbekistan and Kazakhstan, formerly part of the Soviet Union. Most of the water in the Aral Sea came from two rivers, the Amu Dar'ya and the Syr Dar'ya. In the 1960s a huge public works project diverted river water to irrigate enormous fields of rice and cotton. Within 20 years the flow of water into the Aral Sea completely stopped. By 1990 more than half of the Aral Sea's normal volume of water was gone, and the salinity of the remaining water had tripled.

In 1960 about 60,000 people were employed in the fishing industry, which was completely eliminated by the early 1980s. The exposed sea bed is an arid wasteland where winds kick up thousands of tons of dust laden with salt and pesticides, decreasing air quality and crop yields in the region.

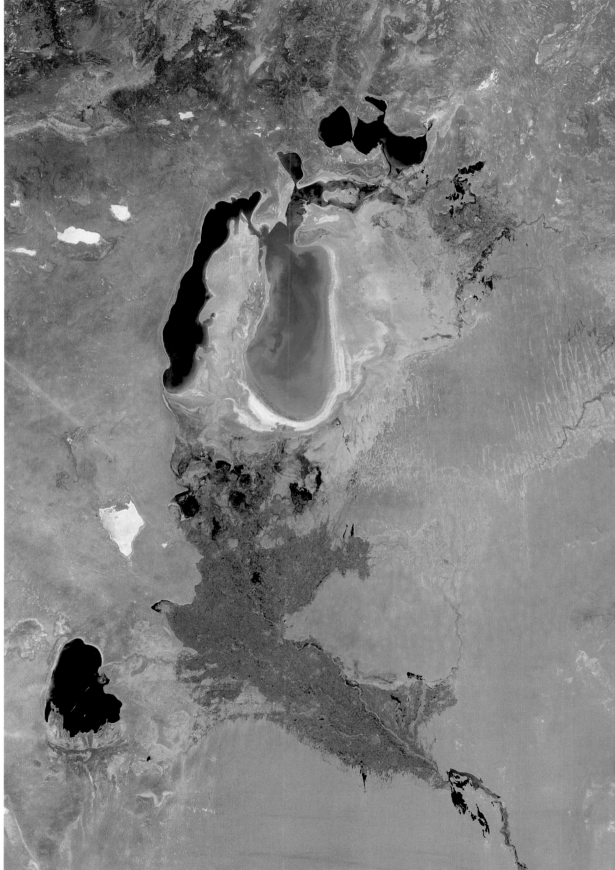

Fluctuating Lake

LAKE CHAD 1968 (LEFT)
1972 (TOP RIGHT)
2002 (LOWER RIGHT)

Lake Chad lies in central Africa, surrounded by the nations of Chad, Nigeria, Niger and Cameroon. It was once the world's sixth-largest lake in area, but the water has never been deeper than about 25 feet (8 m) on the flat lake bed. The lake's volume often triples during the wet season in November, but changes have been noticed over longer periods as well. In the 13th century Lake Chad was high enough to overflow into a neighboring depression to the east. This situation recurred in the early 1960s, but the lake's level has been lower since then. Three different views of Lake Chad are shown here. On the left is a photograph taken at an angle, from the north. On the right are two digital images acquired more recently from directly above the lake.

The photograph on the left, taken by astronauts on the Apollo 7 mission in October 1968, shows Lake Chad covering more than 7,800 square miles (20,000 sq km). By the time the upper-right image was collected by the Landsat 1 satellite in 1972, water level in Lake Chad had fallen significantly. By 1990, years of drought had reduced the lake's area to less than 1,200 square miles (3,000 sq km). The lower-right image shows Lake Chad in 2002. Former lake boundaries are visible, as are ancient sand dunes. The bright red areas are covered by wetland vegetation. Because the water table still lies near the surface, vegetation covers much of the lake's former surface. To the southeast, the effects of irrigation are visible along the Chari River.

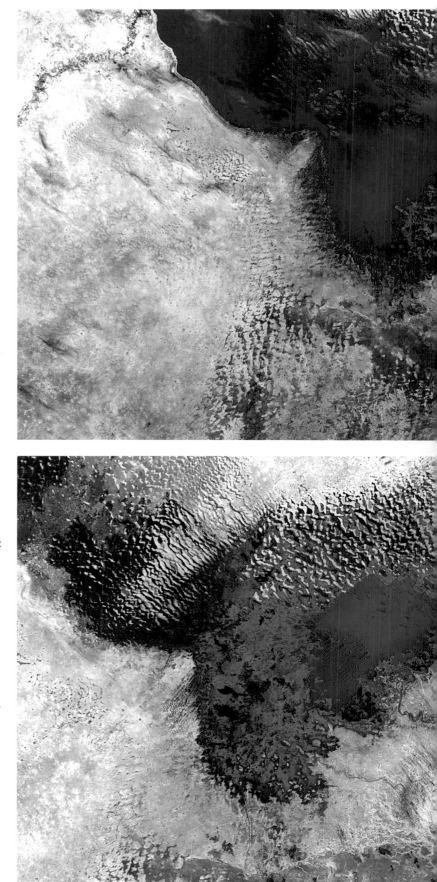

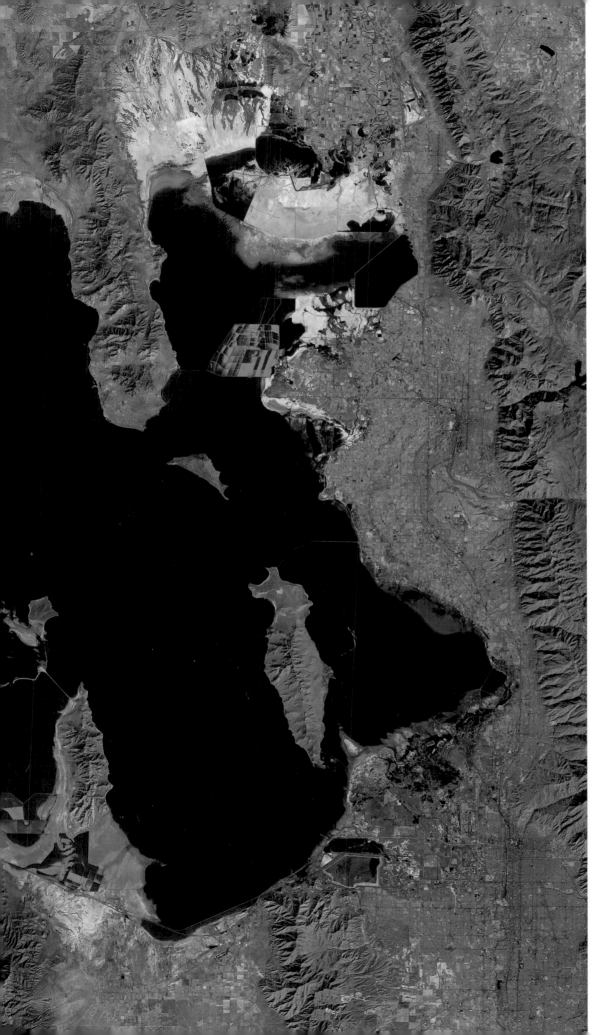

IMAGE ON P. 179

railroad causeway

Great Salt Lake

flooded area

LOWER RIGHT IMAGE

Salt Lake City

N

50 Km

Fluctuating Lake

GREAT SALT LAKE, 1989 (LEFT)

1972 (BELOW)

The largest body of salt water in North America, the Great Salt Lake's water is five times as salty as the ocean's.

The image below was acquired by the Landsat 1 satellite on September 13, 1972, when the water level in Great Salt Lake was very low. Since that time, rising waters have threatened water treatment plants, the airport and other areas of Salt Lake City. The Landsat 7 image on the left from June 1989 shows the lake and flooded areas to the west. Visible in the satellite image is a railroad causeway that cuts across the lake. Made of solid rock and dirt, it prevented water from flowing to the north until 1984, when a 300-foot-long (90 m) opening was created in the causeway.

In 1986 a system of pumps was built to move water to the west. Salty water is pumped out of the north arm of Great Salt Lake and into a canal, which carried the water about 4 miles (2.5 km) to an evaporation pond. The pumping can resume whenever necessary.

Tsunami in the Indian Ocean

KALUTRA, SRI LANKA (ABOVE)

BANDA ACEH, INDONESIA (OPPOSITE PAGE)

On December 26, 2004, an earthquake occurred off the coast of the island of Sumatra, Indonesia. It was one of the largest seismic events ever recorded, and it created a powerful tsunami that swept across the Indian Ocean to Africa. Waves of water more than 50 feet (15 m) high washed ashore, destroying buildings and sweeping far inland. Approximately 200,000 people lost their lives.

The QuickBird satellite was passing over Sri Lanka when the tsunami impacted there. The image above shows water flowing out after the wave flooded coastal areas.

On the opposite page are two images from the IKONOS satellite showing the devastation in Indonesia. The left image shows the city of Banda Aceh and surrounding area on January 10, 2003. The far right image was acquired on December 29, 2004, three days after the tsunami. The city was devastated and much of the area remains flooded.

Changing Seasons

BERING STRAIT, AUGUST (ABOVE)
MAY (RIGHT)

Only about 54 miles (90 km) wide and 100 feet (30 m) deep, the Bering Strait is all that separates North America from Asia. The Seward Peninsula of Alaska lies to the east. The Diomede Islands are in the middle of the strait; one island is U.S. territory, the other is part of Russia. The climate around the Bering Strait varies considerably. Snow cover can persist for eight to nine months on the land at either end of the strait, while rainfall is common in the south.

The image above was collected in summer (August 18, 2000) by the MISR sensor on the Terra satellite. The image to the right, acquired by the MODIS sensor on Terra, shows the Bering Strait as temperatures were warming on May 13, 2002. Ice is slowly melting between the strait and St. Lawrence Island to the south. Areas of open water are surrounded by pack ice in the Chukchi Sea to the north of the strait. Floating ice south of the Bering Strait is usually 4–5 feet (1.2–1.5 m) thick every winter.

Dynamic Ice

CASPIAN SEA (ABOVE)

The Caspian Sea is the largest inland body of water on Earth. Lying in central Asia between Russia, Kazakhstan, Turkmenistan, Azerbaijan and Iran, it covers an area larger than Japan. The lake is over 3,000 feet (900 m) deep in places. The northern part of the Caspian Sea freezes in winter. Small icebergs float south before melting. This image, collected by the MODIS sensor on the Aqua satellite on December 18, 2000, shows new ice freezing. The border of Russia and Kazakhstan has been drawn on the image. The frozen Volga River delta is visible on the Russian side of the border. The Volga Delta can also be seen on page 131.

Fall Foliage

VERMONT (OPPOSITE PAGE)
JUNE (TOP)
OCTOBER (BOTTOM)

As the leaves of deciduous trees lose their chlorophyll in autumn, a brilliant display of colors begins. The two images on the opposite page were acquired by the Landsat 7 satellite in June and October 2002. Both show northern Vermont in visible wavelengths, giving these images approximately the same colors we see with our eyes. Maple, birch and beech trees produce bright reds, oranges and yellows. Trees at higher elevation change color first, and the varied topography enhances the mixture of possible colors.

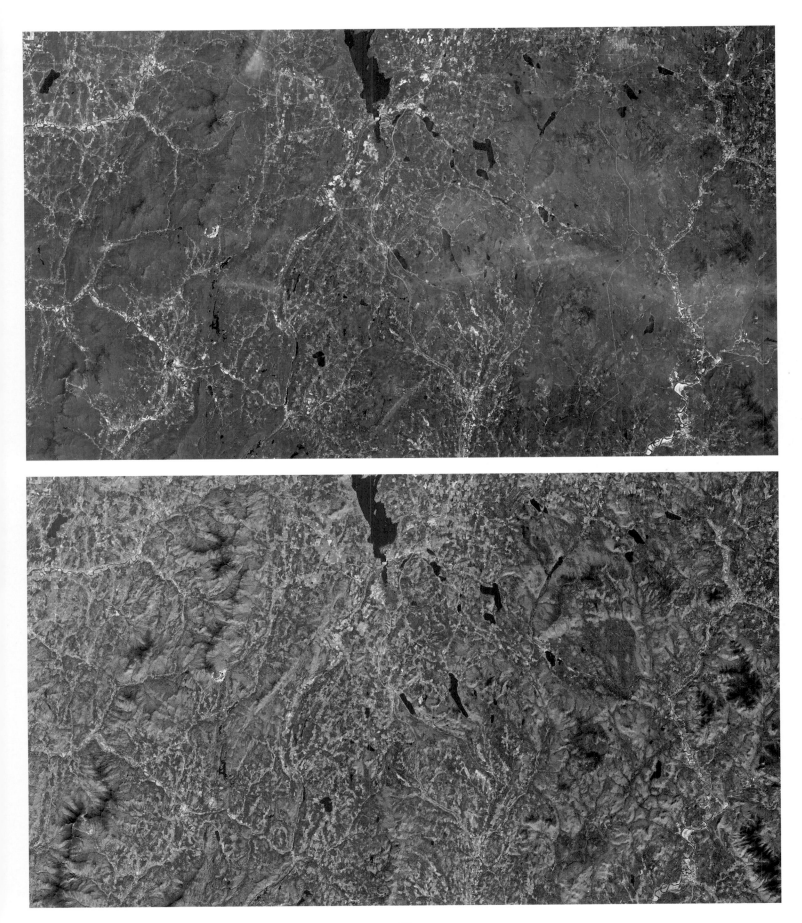

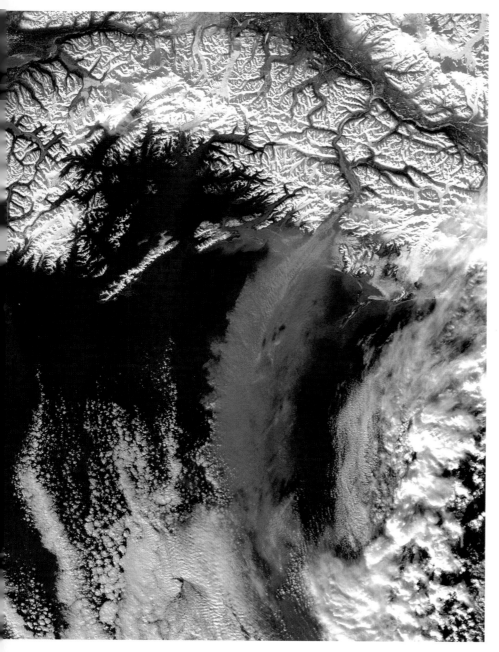

Dust Storms

CENTRAL AFRICA (RIGHT)

In desert areas, dust storms carry tons of dust and sand into the atmosphere. Looking southwest in northern Chad, the photograph on the right was taken by Space Shuttle astronauts in October 1988. It shows patterns of blowing sand and dust created as winds were forced to squeeze between the two dark-colored areas of higher elevation — the Ennedi Plateau to the left, and the Tibesti Mountains to the right. The Ennedi Plateau, made of sandstone, is being slowly eroded by the blowing wind. The volcanic Tibesti Mountains, the highest peaks in the Sahara Desert, are more resistant to erosion. Dust is visible as the lighter-colored plumes rising into the atmosphere. Lake Chad is just visible near the horizon.

COPPER RIVER VALLEY, ALASKA (LEFT)

This image, collected by the MODIS sensor on the Aqua satellite on November 5, 2005, shows a dark gray plume of airborne dust off the coast of Alaska south of Prince William Sound. Large amounts of fine glacial silt are deposited along the Copper River. When this image was acquired the river was relatively dry, allowing strong winds to carry fine particles down the valley and out over the ocean. The plume extends more than 150 miles (240 km) to the south.

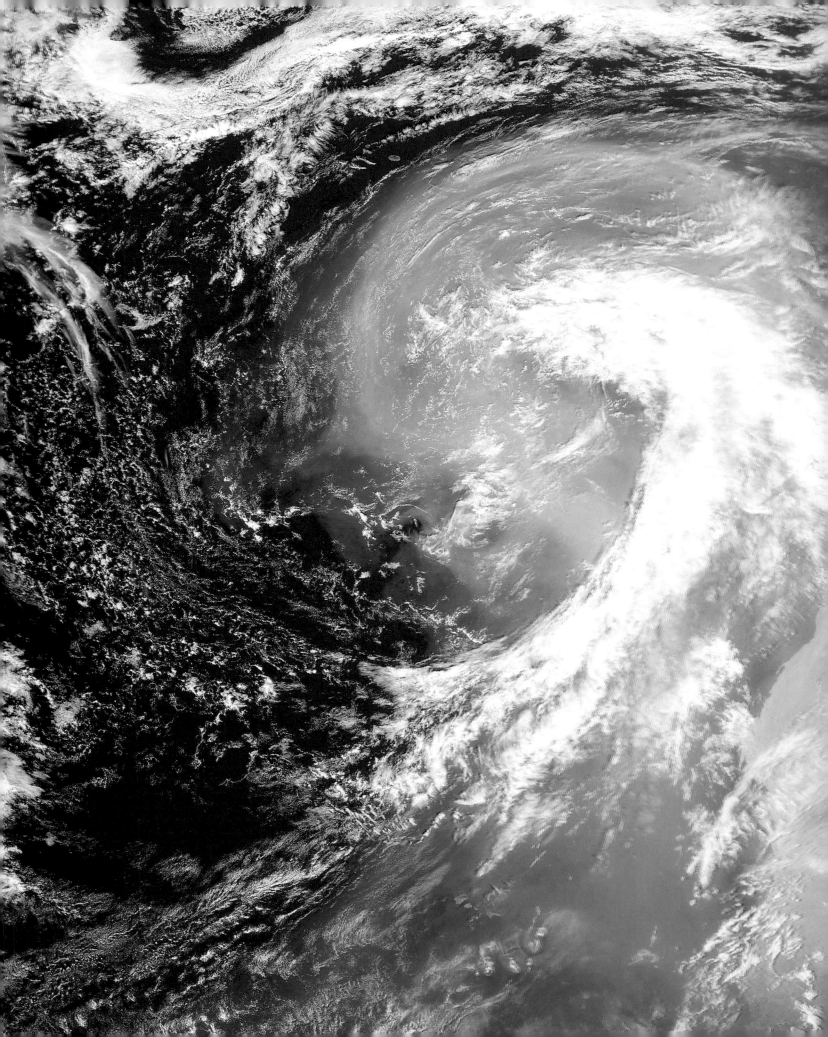

Dust from Africa

OVER THE ATLANTIC

Wind patterns can blow large clouds of
dust from Africa across the Atlantic to
North America. This image, aquired by
the MODIS sensor on the Terra satellite
on June 30, 2002, shows a weather
system that has captured a large plume
of dust from the Sahara Desert. After it
moved over the Canary Islands, the
plume took about a week to cross the
Atlantic. This transfer of dust also moves
bacteria and fungi that may damage reef
ecosystems in the Caribbean. African
dust reaching North America may also
be responsible for respiratory problems.

Urban Growth

DALLAS, 1976 (ABOVE)
2001 (OPPOSITE PAGE, TOP)

The metropolitan area of Dallas, Texas, is seen in these Landsat images. The image above, acquired by the MSS sensor on the Landsat 2 satellite on September 9, 1976, shows Dallas and Forth Worth, which are about 33 miles (53 km) apart. The top image on the opposite page, taken by ETM+ sensor on the Landsat 7 satellite on April 16, 2001, shows the extent of urban growth that has occurred in only 25 years.

In the 1850s Dallas was a small riverbank settlement. In the early 1900s the cotton business thrived around Dallas, and then oil was discovered nearby in the 1930s. The city exploded in size after World War II. Dallas/Fort Worth International Airport opened in 1973, fueling suburban growth between the two urban centers. About 4.5 million people lived in the metropolitan area in 2000.

RIYADH 1972 (OPPOSITE PAGE, BOTTOM LEFT)
2000 (OPPOSITE PAGE, BOTTOM RIGHT)

Riyadh is the largest city and capital of Saudi Arabia. By the early 19th century the city was the political center of this part of the Arabian Peninsula, and it has been the capital of Saudi Arabia since the kingdom was unified in 1932. The world's largest petroleum reserves were discovered in Saudi Arabia in the 1930s. As a capital city, Riyadh grew quickly through the 20th century.

The image on the lower left, acquired by the Landsat 1 satellite in 1972, shows Riyadh when it had a population of about 500,000 and roughly two-thirds of the people of Saudi Arabia lived outside cities. During the next 30 years people flooded into cities, and Riyadh quadrupled in size. The image on the far right was acquired by the ASTER sensor on the Terra satellite on August 24, 2000, when Riyadh had a population of almost 2 million.

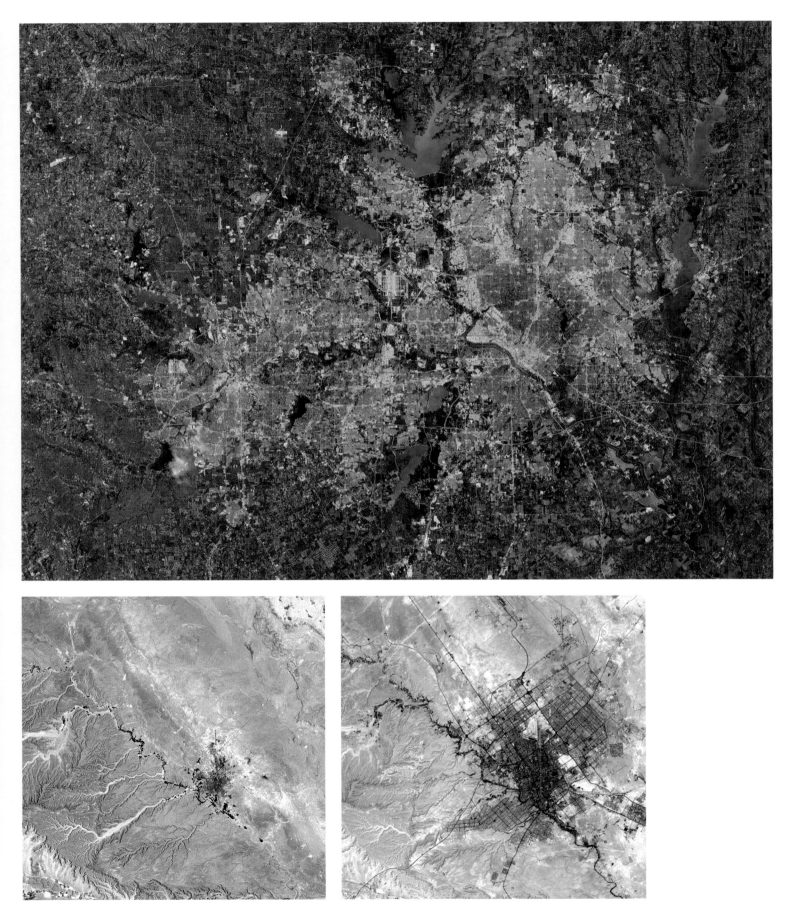

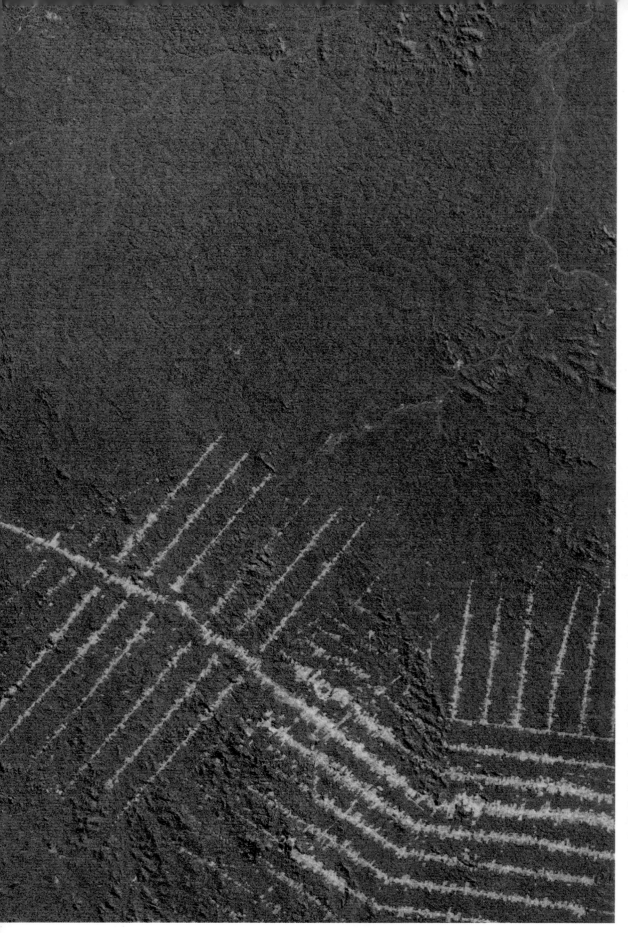

Deforestation

RONDONIA, BRAZIL, 1976 (LEFT)
2001 (RIGHT)

Humans have profound effects on the Earth's forested areas. In the state of Rondônia, Brazil, people have cleared the forest to make room for agriculture and logging. Rondônia is on the frontier between Brazil and Bolivia. In the late 1800s rubber was exported from this area, but the rubber industry collapsed when synthetic material became available. In the late 1900s activity in the area picked up again, and more land than before was cleared to make room for farming, logging and ranching. These images show an area in eastern Rondônia near the confluence of the Jioaraná and Jarú rivers. The main road cutting across the image continues for about 160 miles (260 km) northwest to Porto Velho, the state capital.

These two satellite images show the same area 25 years apart. The image on the left, acquired by the Landsat 1 satellite on June 21, 1976, shows the deforestation in its early stages, with a distinctive "fish bone" pattern of penetration into the forest. The image on the right was acquired by the Landsat 7 satellite on August 11, 2001. New roads cut through the Amazon basin in the 1980s, providing access to this area. As the roads are constructed, land on each side is cleared of trees. Forest cover appears red in these infrared views, each covering an area about 30 miles (50 km) wide.

Elbe River

BEFORE FLOODING (TOP LEFT)
DURING FLOODING (LOWER LEFT)

In August 2002 many rivers in central Europe were flooded by unusually heavy rainfall. The Elbe River flooded its banks, reaching levels not seen since 1845. The floods killed more than a hundred people and caused $20 billion in damage to Germany, Austria, Hungary, Russia and the Czech Republic. These infrared images from the Landsat 7 satellite show the area around Wittenberg, Germany, before and during flooding. The areas covered by water are blue, and vegetation appears bright green. Prague, about 200 miles (320 km) upstream, experienced some of the worst damage. Its zoo was badly damaged. A seal named Gaston survived being swept hundreds of miles downstream, but he later died.

Recovering from a Storm

NEW ORLEANS (RIGHT)

The city of New Orleans was devastated after Hurricane Katrina in 2005. This image from the QuickBird satellite from August 31, 2005, shows widespread flooding two days after the storm. The darker color of the streets in the lower half indicates areas under water. The Superdome, the round building near the center, was stripped of much of its white roofing. The city's historic French Quarter, which did not experience flooding, is visible in the upper left. Katrina was responsible for more than 1,000 deaths and caused massive property damage. An image showing an overview of the area can be seen on page 207.

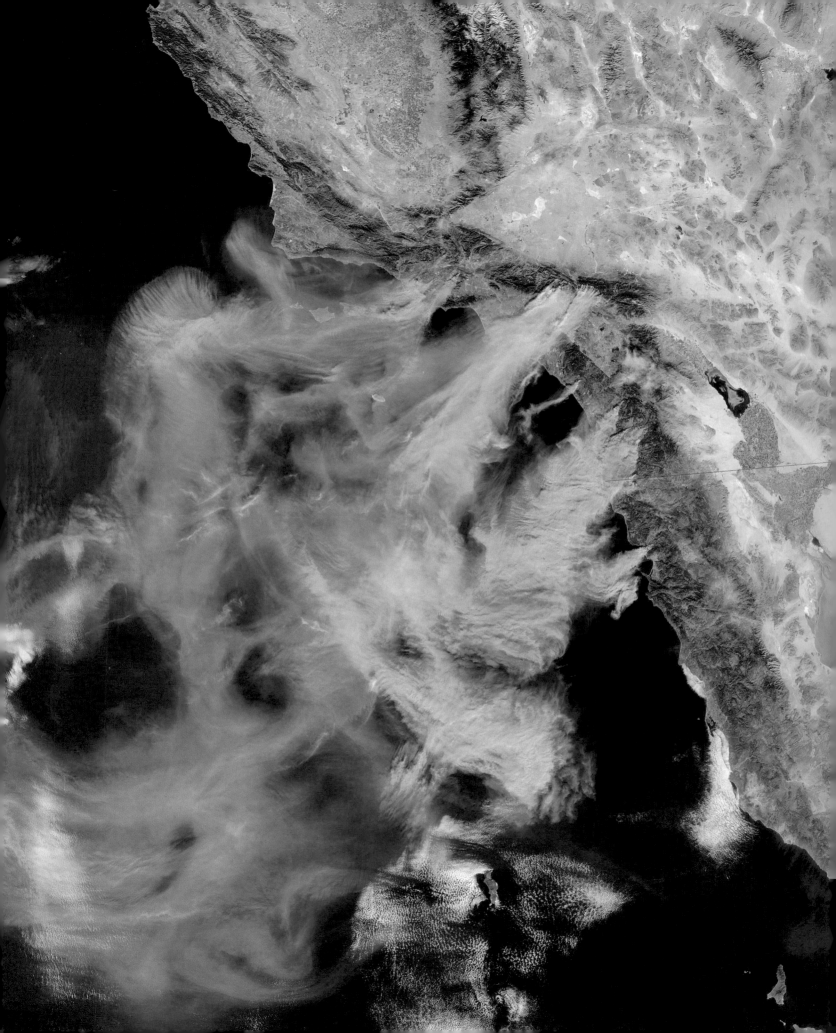

Burning Forests

CALIFORNIA (LEFT)

Wildfires, started by natural forces or by people, often endanger cities. This image shows southern California under assault by some of the largest fires that the area has ever seen. Winds from the Mojave Desert, known as the Santa Ana winds, often carry fires toward populated areas. These fires began in October 2003, and within a few days they were raging through the area. Several hundred homes were destroyed, and 13 people lost their lives.

At least five groups of fires can be seen burning in this image, which was acquired on October 26, 2003, by the MODIS sensor on the Terra satellite. Huge clouds of smoke are blowing over San Diego and Los Angeles and spreading across the Pacific Ocean.

SYDNEY (ABOVE LEFT)

This image, acquired by the MODIS sensor on the Terra satellite on December 25, 2001, shows plumes of smoke billowing over Sydney, Australia. Forests appear dark green. MODIS can detect heat emitted from the Earth's surface to identify fire locations.

Sydney, lying between the ocean and forested mountains, is a dry area susceptible to fires. Each year fires break out in November and December. Many dry years created a dangerous situation at the end of the 20th century, when the worst drought in a century made forests especially vulnerable. In 2001 and 2002 fires destroyed neighborhoods built on the hills outside central Sydney and came within 15 miles (9 km) of the city. The next year another round of severe fires swept through the hills.

BORNEO (ABOVE RIGHT)

More than 12 million people live on the island of Borneo. The small state of Brunei is on the northwest coast, and the rest of the island is divided between Indonesia and Malaysia. The Indonesian part of Borneo is called Kalimantan.

Every year fires burn in the tropical forests that cover most of the island. Many fires are deliberately set to clear land for agriculture. The smoke pollutes the air, disrupts air traffic and causes health problems.

During 2002 Borneo experienced more than its usual number of fires. The image above was acquired by the MODIS sensor on the Terra satellite on August 19, 2002. Dozens of fires are creating a thick layer of haze. Large fires near the Kapuas River are producing long trails of smoke visible in the left side of the image. A huge cloud of smoke hangs over the southeast, trapped against the central highlands by prevailing winds. Smoke from these fires disrupted air travel as far away as Singapore.

Sources of Remote Sensing Imagery

National Oceanic and Atmospheric Administration

National Environmental Satellite, Data, and Information Service

Geostationary Satellite Server
This server provides access to images from NOAA GOES geostationary satellites. These images are updated frequently and are used to track weather systems. The website also contains links to images from other geostationary satellites.
www.goes.noaa.gov

Operational Significant Event Imagery
The Operational Significant Event Imagery team produces high-resolution imagery of significant environmental events. The website outlines events captured in satellite imagery and provides a link to each image.
www.osei.noaa.gov

NASA Johnson Space Center

Earth Sciences and Image Analysis Laboratory
The Earth Sciences and Image Analysis Laboratory at NASA's Johnson Space Center maintains the collection of photographs taken by U.S. astronauts. Since 1961 astronauts have taken more than 500,000 photographs of the Earth from orbit. Astronauts on the International Space Station continue to take new photographs. The collection can be accessed at no cost through the websites listed below.

Gateway to Astronaut Photography of Earth
This website provides access to the full range of astronaut photographs with location data and descriptions.
eol.jsc.nasa.gov

Earth From Space
This website provides access to selected astronaut photographs, which can be searched by geographical locations or themes. The Earth From Space website went online in 1995 and is not affiliated with this book.
earth.jsc.nasa.gov/sseop/efs

NASA Jet Propulsion Laboratory

ASTER Project
ASTER (Advanced Spaceborne Thermal Emission and Reflection Radiometer) is a remote sensing instrument on the Terra satellite. The website contains a description of the instrument and a growing selection of recent images.
asterweb.jpl.nasa.gov

MISR Project
MISR (Multi-Angle Scanning Radiometer) is a remote sensing instrument on the Terra satellite. The website contains a description of the instrument and selections of images.
www-misr.jpl.nasa.gov

Planetary Photojournal
This website offers access to a wide range of satellite images of the Earth and other planets.
photojournal.jpl.nasa.gov

NASA Goddard Space Flight Center

MODIS Project
MODIS (Moderate Resolution Imaging Spectrometer) is a remote sensing instrument on the Terra and Aqua satellites. The website contains a description of the instrument and selections of recent images. The gallery is updated daily with new images.
modis.gsfc.nasa.gov

Visualization Analysis Laboratory
The lab produces visualizations of remote sensing data for science, education and outreach. These images were designed for use in publications and for display in public settings. The website includes a link to global images produced by the lab.
val.gsfc.nasa.gov

MODIS Rapid Response System
This server provides access to imagery acquired by the MODIS sensors on the Terra and Aqua satellites. An extensive image gallery is included. Each day a new image is featured. Images can be searched by date and location and downloaded at three different resolutions.
rapidfire.sci.gsfc.nasa.gov

Earth Observatory
This NASA website provides remote sensing images from a wide range of orbiting platforms. Images are grouped into broad themes, with a focus on Earth's climate and environmental change. The site is updated daily with new images. Special topics are covered in depth with stories about remote sensing applications.
earthobservatory.nasa.gov

Visible Earth
Visible Earth is a NASA website that provides a consistently updated, central point of access to NASA's Earth science-related images, animations and data visualizations. These images are freely available to the public. Credit and caption information is provided for all materials.
visibleearth.nasa.gov

U.S. Geological Survey

EROS Data Center
The Earth Resources Observation Systems (EROS) Data Center is a data management and research field center for the U.S. Geological Survey's National Mapping Division. The EROS Data Center serves as the primary public source of data from many remote sensing satellites operated by the U.S. government, including Landsat, Terra and Aqua satellites.

Global Visualization Viewer
This web-based application allows visitors to search a global collection of Landsat 7 and ASTER data. Images can be identified by geographic coordinates and ordered for download.
glovis.usgs.gov

Earth Explorer
This website allows visitors to access a wide range of remote sensing data. This includes Landsat, Terra, Aqua, EO-1, air photos and declassified reconnaissance programs. Orders can be placed for delivery of data.
earthexplorer.usgs.gov

Landsat Image Gallery
This gallery is updated frequently to demonstrate images acquired by the Landsat 7 satellite. Links are included to collections of especially interesting images and images identifying changes over time.
landsat.usgs.gov/gallery

GeoEye
GeoEye is a commercial company that operates a variety of remote sensing satellites. The company was formed in 2006 with the merger of ORBIMAGE and Space Imaging. ORBIMAGE launched the OrbView-2 satellite in 1997 and the OrbView-3 high-resolution imaging satellite in 2003. Space Imaging operated the IKONOS satellite. The website includes links to images from the OrbView and IKONOS satellites and other data sources.
www.geoeye.com

DigitalGlobe
DigitalGlobe operates the QuickBird satellite, providing high-resolution imagery for GIS and mapping applications. The website includes an extensive gallery of QuickBird images. Details about using QuickBird data are included and image products can be ordered.
www.digitalglobe.com

ImageSat International
ImageSat International N.V. is a commercial provider of high-resolution satellite imagery. Its Earth Resources Observation Satellite A (EROS A) was launched in 2000. EROS B was launched in 2006. The website contains an extensive collection of full-resolution imagery.
www.imagesatintl.com

MDA Geospatial Services
MDA Geospatial Services distributes remote sensing data from platforms such as RADARSAT, Landsat, Envisat, Quickbird, and IKONOS. MDA Federal, formerly Earth Satellite Corporation, provides satellite images and derived products for projects that focus on resource management and monitoring of the environment. The website includes a gallery with examples of a variety of satellite images.
gs.mdacorporation.com
www.mdafederal.com

Spot Image
Spot Image distributes data from SPOT 5 and several other remote sensing satellites. Both high-resolution images and wide-area environmental data are collected by the SPOT satellites. Spot Image operates globally, providing local-, regional- and continental-scale images acquired by a range of sensors.
www.spotimage.fr

Indian Space Research Organisation
The Indian Space Research Organisation (ISRO) is a governmental organization that develops space technology and applications. It has operated the IRS series of satellites designed to observe the Earth's environment. Since 2003 ISRO has managed Resourcesat-1, a satellite with enhanced capabilities.
www.isro.org

Research and Development Center ScanEx

Research and Development Center ScanEx is a private company providing integrated solutions for satellite remote sensing applications. ScanEx processes data from a range of remote sensing satellites and provides technologies for downloading images. The website includes an image gallery.

www.scanex.ru

Scientific Research Center Planeta

This web server is a joint project of the Scientific Research Center Planeta and the Russian Space Research Institute. It provides information about the Resurs-01, Meteor and Okean-01 satellites. Examples of images are included and access to imagery catalogs is provided.

sputnik.infospace.ru

Sovinformsputnik

SOVINFORMSPUTNIK distributes data from the TK-350, KVR-1000, and other camera systems that operate on the Cosmos series of satellites. These photographs provide high-resolution data over widely distributed points across the globe.

www.sovinformsputnik.com

European Space Agency
Envisat Mission

The European Space Agency's Envisat is a polar-orbiting Earth observation satellite. The instruments on Envisat were designed to observe environmental changes on the Earth's surface and in the atmosphere. The website includes links to images from the MERIS sensor and other instruments on Envisat.

envisat.esa.int

EUMETSAT

EUMETSAT is an organization of European nations that operates the Meteosat weather satellites. EUMETSAT is responsible for the launch and operation of the satellites and for delivering satellite data. The website contains links to current images.

www.eumetsat.int

Satellite Imaging

Satellite Imaging Corporation processes and distributes remote sensing data from a wide range of satellites and airborne platforms. The company provides imagery for a variety of applications. The website includes a gallery with examples of images.

www.satimagingcorp.com

EarthKAM

The EarthKAM is a digital camera on the International Space Station that provides images of wide areas of the Earth. School students have taken thousands of photographs by controlling the camera from their classrooms. The site provides access to images and details about the program.

earthkam.ucsd.edu

Arizona State University
Geological Remote Sensing Laboratory
Urban Environmental Monitoring Project

This site contains links to an extensive collection of images showing urban areas. These images were acquired by the ASTER sensor on the Terra satellite. Cities throughout the world can be selected using maps or continental lists. Images from different dates for each city can be downloaded.

elwood.la.asu.edu/grsl/UEM

University of Maryland
Global Land Cover Facility

The Global Land Cover Facility is a distribution point for processed satellite imagery and derived remote sensing products. The web page includes a link to the Earth Science Data Interface, where a wide range of digital remote sensing data can be searched and downloaded at no charge. Thousands of Landsat images and data from other spacecraft are available.

glcfapp.umiacs.umd.edu

Smithsonian Institution
National Air and Space Museum
Center for Earth and Planetary Studies

The office maintains several image collections that are available as references to researchers in the scientific community. It houses a Regional Planetary Image Facility, which is a NASA-supported reference library containing image data obtained from planetary missions. Earth-looking imagery is available from early manned missions and the Space Shuttle photograph collection.

www.nasm.si.edu/ceps

Earth From Space exhibit

This museum exhibit, based on the Earth From Space book, is a collaboration of the Smithsonian Institution Traveling Exhibitions Service and the National Air and Space Museum. The exhibit opened at the National Air and Space Museum in 2006. The web site includes more information about the exhibit, details on many of the images in this book, and educational materials.

www.earthfromspace.si.edu

Remote Sensing Satellites

Terra
Launched: 1999
Orbit: polar, Sun-synchronous, 702 km altitude
Operated by: National Aeronautics and Space
   Administration

Imaging sensors:
MODIS (Moderate Resolution Imaging Spectroradiometer)
Spectral channels: 36 channels in visible and infrared
   wavelengths
Spatial resolution: 250 m in visible, 500 m in near-
   infrared, 1000 m in mid-infrared

ASTER (Advanced Spaceborne Thermal Emission and
   Reflection Radiometer)
Spectral channels: 14 channels, 4 in visible and near-
   infrared, 6 in shortwave infrared, and 5 in thermal-
   infrared
Spatial resolution: 15–90 m

MISR (Multi-angle Imaging SpectroRadiometer)
Spectral channels: 4, with 9 cameras at different angles
Spatial resolution: 275–550 m

CERES (Clouds and the Earth's Radiant Energy System)
Spectral channels: 3, measures reflected and emitted
   radiation
Spatial resolution: wide-area sensor

MOPITT (Measurements Of Pollution In The Troposphere)
Channels: 3, detects carbon dioxide and methane in the
   atmosphere.
Spatial resolution: 22 km

Aqua
Launched: 2002
Orbit: polar, Sun-synchronous, 702 km altitude
Operated by: National Aeronautics and Space
   Administration

Imaging sensors:
MODIS (Moderate Resolution Imaging Spectroradiometer)
Same details as the MODIS on Terra

CERES (Clouds and the Earth's Radiant Energy System)
Same details as the CERES on Terra

HSB (Humidity Sounder for Brazil)
Spectral channels: 4-channel passive microwave sounder
Spatial resolution: 13.5 km

AIRS (Atmospheric Infrared Sounder)
Spectral channels: 2378 infrared channels for atmospheric
   sounding
Spatial resolution: 13.5 km

AMSR-E (Advanced Microwave Radiometer for EOS)
Spectral channels: 6-channel passive microwave sounder
Spatial resolution: 4–43 km

AMSU (Advanced Microwave Sounding Unit)
Spectral channels: 15-channel passive microwave sounder
Spatial resolution: 40 km

Landsat 5
Launched: 1984
Orbit: polar sun synchronous, 705 km altitude
Operated by: National Aeronautics and Space
   Administration, U.S. Geological Survey
Imaging Sensors:
   TM (Thematic Mapper)

spectral channels: 8 in visible, near IR, and thermal IR
   wavelengths
spatial resolution: 30 m multispectral, 120m thermal IR
MSS (MultiSpectral Scanner)
   spectral channels: 4 in visible and near IR wavelengths
   spatial resolution: 80 m

Landsat 7
Launched: 1999, sensor partially malfunctioned 2003
Orbit: polar, Sun-synchronous, 702 km altitude
Operated by: National Aeronautics and Space
   Administration, U.S. Geological Survey
Imaging Sensor:
ETM+ (Enhanced Thematic Mapper Plus)
Spectral channels: 8 in visible, near-infrared and thermal-
   infrared wavelengths
Spatial resolution: 15 m panchromatic, 30 m multispectral,
   60 m thermal-infrared

IKONOS
Launched: 1999
Orbit: Sun-synchronous, 681 km altitude
Operated by: GeoEye
High-resolution sensor:
Channels: 5 in visible and infrared
Spatial resolution: 1 m panchromatic, 4 m multispectral

GEOEYE-1
Launch planned for 2007
Orbit: polar sun synchronous, 684 km altitude
Operated by: GeoEye
High resolution sensor:
Channels: 4
Maximum spatial resolution: 0.41 m

QuickBird
Launched: 2001
Orbit: polar, Sun-synchronous, 450 km altitude
Operated by: DigitalGlobe
High-resolution sensor:
Spectral channels: 4 in visible and near-infrared
   wavelengths
Spatial resolution: 61 cm panchromatic, 2.4 m
   multispectral

WorldView I
Launch planned for 2007
Orbit: polar sun synchronous, 450 km altitude
Operated by: DigitalGlobe
High resolution sensor:
Channels: 1
Maximum spatial resolution: 0.45 m

OrbView-2
Launched: 1997
Orbit: polar, Sun-synchronous, 300 km altitude
Operated by: GeoEye
SeaWiFS (Sea viewing Wide Field of view Sensor):
Spectral channels: 8 channels in visible and infrared
   wavelengths
Spatial resolution: 1 km

OrbView-3
Launched: 2003
Orbit: polar, Sun-synchronous, 470 km altitude
Operated by: GeoEye
High-resolution sensor:
Spectral channels: 4 channels in visible and infrared
   wavelengths
Spatial resolution: 1 m panchromatic, 4 m multispectral

EROS A
EROS: Earth Resources Observation Satellite
Launched: 2000
Orbit: Sun-synchronous, 480 km altitude
Operated by: ImageSat International N.V.
High-resolution sensor:
Channels: 1
Spatial resolution: 1.8 m

EROS B
Launched: 2006
Orbit: polar sun synchronous, 500 km altitude
Operated by: ImageSat International N.V.
High resolution sensor:
Channels: 1
Spatial resolution: 0.7 m

EnviSat
Launched: 2002
Orbit: polar, Sun-synchronous, 800 km altitude
Operated by: European Space Agency

Imaging sensors:
MERIS (MEdium Resolution Imaging Spectrometer)
Spectral channels: 15 in visible and infrared wavelengths
Spatial resolution: 300–1200 m

ASAR (Advanced Synthetic Aperture Radar)
   Spectral channels: C-band radar instrument
Spatial resolution: 30–1000 m

SCIMACHY (SCanning Imaging Absorption spectroMeter
   for Atmospheric CHartographY)
Spectral channels: 8 in UV, visible, and infrared for
   atmospheric sounding
Spatial resolution: 300 m

AATSR (Advanced Along Track Scanning Radiometer)
Spectral channels: 7
Spatial resolution: 1000 m

ERS 2
Launched: 1995
Orbit: polar, Sun-synchronous, 780 km altitude
Operated by: European Space Agency

Imaging sensors:
GOME (Global Ozone Monitoring Experiment)
Channels: 4 channels in UV and visible wavelengths
Spatial resolution: 300 m

SAR (Synthetic Aperture Radar)
Spectral channels: C-band radar instrument
Spatial resolution: 25 m

ATSR-2 (Along Track Scanning Radiometer)
Spectral channels: 7
Spatial resolution: 1000 m

GOES 9, GOES 10, GOES 11, GOES 12

GOES: Geostationary Operational Environmental Satellite
GOES 9 launched 1995
GOES 10 launched 1997
GOES 11 launched 2000
GOES 12 launched 2001
Orbit: geostationary, 36,000 km altitude
Operated by: National Oceanic and Atmospheric
Administration

Imaging Sensors:
GOES Imager:
Spectral channels: 5 in visible, infrared and thermal-
infrared wavelengths
Spatial resolution: 1 km visible, 4–8 km infrared

GOES Sounder:
Spectral channels: 19 channels in infrared wavelengths
Spatial resolution: about 30 km

RADARSAT-1

Launched: 1995
Orbit: polar, Sun-synchronous, 798 km altitude
Operated by: Canadian Space Agency
Radar system:
Spectral channels: C-band radar sensor
Spatial resolution: 25 m

SPOT 5

Launched: 2002
Orbit: polar, Sun-synchronous, 832 km altitude
Operated by: SPOT Image

Imaging Sensors:
HRG (High Geometric Resolution) sensor:
Spectral channels: 5 channels in visible and infrared
Spatial resolution: 5 m panchromatic, 10 m multispectral

HRS (High Resolution Stereoscopic) sensor
Spectral channels: 1
Spatial resolution: 10 m

VEGETATION
Spectral channels: 4 channels in visible and infrared
Spatial resolution: 1 km

SPOT 4

Launched: 1998
Orbit: polar, Sun-synchronous, 832 km altitude
Operated by: SPOT Image

Imaging Sensors:
HRVIR (High Visible InfraRed) sensor:
Spectral channels: 4 channels in visible and infrared
Spatial resolution: 10 m panchromatic, 20 m multispectral

VEGETATION
Same details as the VEGETATION sensor on SPOT 5

Resourcesat-1

Launched: 2003
Orbit: polar, Sun-synchronous, 817 km altitude
Operated by: Indian Space Research Organisation

Imaging Sensors:
LISS 4 (Linear Imaging Self-Scanner)
Spectral channels: 3 in visible and near-infrared
wavelengths
Spatial resolution: 5.8 m

LISS 3 (Linear Imaging Self-Scanner)
Spectral channels: 3 in visible and infrared wavelengths
Spatial resolution: 24 m

AWiFS (Advanced Wide Field Sensor)
Channels: 4
Spatial resolution: 188 m
Surface coverage: 800 km swath

IRS-1D

Launched: 1996
Orbit: polar, Sun-synchronous, 817 km altitude
Operated by: Indian Space Research Organisation

Imaging Sensors:
LISS 3 (Linear Imaging Self-Scanner)
Same details as the LISS3 on Resourcesat-1

PAN sensor
Spectral channels: 1
Spatial resolution: 5–8 m

WIFS (Wide Field Sensor)
Channels: 4
Spatial resolution: 188 m

METEOSAT 6, METEOSAT 7

Launched: 1993, 1997
Orbit: geostationary, 36,000 km altitude
Operated by: EUMETSAT

Imaging Sensor:
VISSR (Visible Infrared Spin Scan Radiometer)
Channels: 3: visible, infrared and thermal-infrared
Spatial resolution: 2500–5000 m

METEOSAT 8

Launched 2002
Orbit: geostationary, 36,000 km altitude
Operated by: EUMETSAT

Imaging Sensor:
SEVIRI (Spinning Enhanced Visible and Infrared Imager)
Spectral channels: 12 in visible, infrared, and thermal-
infrared
Spatial resolution: 1 km visible, 3 km infrared

NOAA 16, NOAA 17

Launched: 2000, 2002
Orbit: polar, Sun-synchronous, 850 km (NOAA 16), 812 km
(NOAA 17) altitude
Operated by: National Oceanic and Atmospheric
Administration as part of the Polar Orbiting
Environmental System (POES)

Imaging Sensors:
AVHRR (Advanced Very High Resolution Radiometer)
Spectral channels: 5 in visible and infrared wavelengths
Spatial resolution: 1100 m

AMSU (Advanced Microwave Sounding Unit)
Spectral channels: 20 channel passive microwave sensor
Spatial resolution: 44 km

GMS 5

GMS: Geostationary Meteorological Satellite
Launched: 1994
Orbit: geostationary, 36,000 km altitude
Operated by: Japan Meteorological Agency

Imaging Sensor:
VISSR (Visible and Infrared Spin Scan Radiometer)
Spectral channels: 4 in visible and infrared wavelengths
Spatial resolution: 1.25 km visible, 5 km infrared

DMSP

DMSP: Defense Meteorological Satellite Program
Launched: five operational satellites, launched 1994–2003
Orbit: polar, Sun-synchronous, 830 km altitude
Operated by: U.S. Air Force Weather Agency

Imaging Sensors:
OLS (Operational Linescan System)
Spectral channels: 2, visible and thermal-infrared
wavelengths
Spatial resolution: .5 km

SSM/I (Microwave Imager)

Spectral channels: 4 passive microwave frequencies
Spatial resolution: 13–69 km

TRMM

TRMM – Tropical Rainfall Measuring Mission
Launched: 1997
Orbit: polar, 350 km altitude
Operated by: National Aeronautics and Space
Administration

Imaging Sensors:
Precipitation Radar
Spectral channels: 1, detects radar echo from falling
raindrops
Spatial resolution: 250 km

TMI (TRMM Microwave Imager)
Spectral channels: 5 passive microwave frequencies
Spatial resolution: 6–35 km

VIRS (Visible and Infrared Scanner)
Spectral channels: 5, in visible and mid IR wavelengths
Spatial resolution: 2 km

RESURS-O1 3, RESURS-O1 4

Launched: 1994, 1998
Orbit: polar, Sun-synchronous, 678 km altitude
Operated by: Scientific Research Center Planeta

Imaging Sensors:
MSU-E
Spectral channels: 3 in visible and near-infrared
wavelengths
Spatial resolution: 45 m

MSU-SK
Spectral channels: 5 in visible, infrared and thermal-
infrared wavelengths
Spatial resolution: 160 m, 600 m thermal-infrared

RESURS-DK1

Launched: 2004
Orbit: elliptical, 70 degree inclination
Imaging sensor:
Spectral channels: 1 panchromatic channel
Spatial resolution: 1 m

Kometa Mapping system

Launched: occasional launches on Cosmos satellites
Orbit: circular, 220 km altitude
Operated by: Russian Aviation and Space Agency

Camera Systems:
TK-350
Spectral cahnnels: 1
Spatial resolution: 20 m

KVR-1000
Spectral channels: 1
Spatial resolution: 10 m

Further Reading

Aherns, C. Donald. *Meteorology Today*. St. Paul: West Publishing, 1988.

Allen, Thomas B. *America From Space*. Toronto: Firefly Books, 1998.

Bodechtel, Johann. *The Earth from Space*. New York: Arco Reprints, 1974.

Bourne, Larry S., ed. *Internal Structure of the City*. Oxford: Oxford University Press, 1982.

Burenhult, Göran, ed. *Old World Civilizations: The Rise of Cities and States*. New York: HarperCollins, 1994.

Chiras, Daniel D. *Environmental Science: A Framework for Decision Making*. San Francisco: Benjamin Cummings, 1998.

Earle, Sylvia A. *Atlas of the Ocean*. Washington DC: National Geographic, 2001.

Glassner, Martin Ira and Harm J. de Blij. *Systematic Political Geography*. New York: John Wiley and Sons, 1989.

Hamblin, W. Kenneth. *The Earth's Dynamic Systems*. Minneapolis: Burgess Publishing, 1985.

Hanson, Susan, ed. *The Geography of Urban Transportation*. New York: Guilford Press, 1986.

Harris, Nathaniel. *Mapping the World: Maps and Their History*. San Diego: Thunder Bay Press, 2002.

Hughes, Robert and German Space Center. *Planet Earth*. New York: Alfred A. Knopf, 2002.

Jensen, J.R. *Remote Sensing of the Environment: An Earth Resource Perspective*. Saddle River NJ: Prentice Hall, 2000.

Krebs, Charles J. *Ecology*. New York: HarperCollins, 1994.

Lauber, Patricia. *Seeing Earth from Space*. London: Orchard Books, 1994.

Levin, Harold L. *The Earth Through Time*. Austin: Holt Rinehart Winston, 1988.

Lillesand, Thomas M. and Ralph W. Kiefer. *Remote Sensing and Image Interpretation*. New York: John Wiley and Sons, 2000.

Luhr, James F., ed. *Earth*. New York: DK Publishing, 2003.

Mueller-Wille, Christopher. *Images of the World*. New York: Rand McNally, 1983.

National Geographic. *Into the Unknown: The Story of Exploration*. Washington DC: National Geographic, 1987.

National Geographic. *Satellite Atlas of the World*. Washington DC: National Geographic, 1998.

Rees, W.G. *Physical Principles of Remote Sensing*. Cambridge: Cambridge University Press, 1996.

Sabins, Floyd F. *Remote Sensing: Principles and Interpretation*. San Francisco: W.H. Freeman, 1996.

Silverman, Diana, Emilio F. Moran, Ronald R. Rindfuss and Paul C. Stern, eds. *People and Pixels: Linking Remote Sensing and Social Science*. Washington DC: National Academy Press, 1998.

Strain, Priscilla and Frederick Engle. *Looking at Earth*. Atlanta: Turner Publishing, 1992.

Turner, B.L. and William B. Meyer, eds. *Changes in Land Use and Land Cover: A Global Perspective*. Cambridge: Cambridge University Press, 1994.

Watters, Thomas R. *Planets: A Smithsonian Guide*. New York: Macmillan, 1995.

Williams, J.E.D. *From Sails to Satellites*. Oxford: Oxford University Press, 1992.

Photo Credits

3   NASA/GSFC

6   National Air and Space Museum

8   NASA/GSFC

10   Clarity of the Atmosphere, Reto Stockli/NASA/GSFC MODIS Science Team

11   Temperature of the Seas, NASA/GSFC

11   Global Cloud Cover, NASA/GSFC ISCCP

12–13   Global-to-Local Scales, NASA/GSFC Visualization Analysis Laboratory

14   Earth Satellite Corporation

16   Space Imaging

18   National Air and Space Museum

19   National Air and Space Museum

20   NASA/LaRC, CERES Projects

21   NASA/LaRC, CERES Projects

22   NASA/UCSD

23   NASA/GSFC

26   National Air and Space Museum

27   Indiana, National Air and Space Museum

27   Indianapolis, National Air and Space Museum

28–29   NASA/GSFC/METI/ERSDAC/JAROS, U.S./Japan ASTER Science Team

30   NASA/GSFC

31   Jacques Descloitres, MODIS Land Rapid Response Team, NASA/GSFC

32   DigitalGlobe

34   DigitalGlobe

35   GeoEye

36   Jesse Allen and Laura Rocchio, NASA Earth Observatory, USGS

37   NASA/JPL

38   ImageSat International

39   Jacques Descloitres, MODIS Land Rapid Response Team, NASA/GSFC

40   DigitalGlobe

41   GeoEye

42   GeoEye

43   GeoEye

44   USGS/EROS Data Center

45   DigitalGlobe

46   USGS/EROS Data Center

48   SeaWiFS Project, NASA/GSFC and GeoEye

49   SeaWiFS Project, NASA/GSFC and GeoEye

50   NASA/GSFC

51   NASA/GSFC

54   Lake Natron, NASA/JSC/Earth Sciences and Image Analysis, photos ISS005-E-2054, 2055

54   Lake Carnegie, USGS/EROS Data Center

55   Robert Simmon, NASA/GSFC Landsat 7 Science Team

56   Jacques Descloitres, MODIS Land Rapid Response Team, NASA/GSFC

57   NASA/GSFC/LaRC/JPL, MISR Team

58   CNES/SPOT Image

59   USGS/EROS Data Center, National Air and Space Museum

60   National Air and Space Museum

61   Jacques Descloitres, MODIS Land Rapid Response Team, NASA/GSFC

62   SeaWiFS Project, NASA/GSFC and GeoEye

64   NASA/JSC/Earth Sciences and Image Analysis, photo STS068-244-093

65   NOAA/NESDIS

66   Jeff Schmaltz, MODIS Rapid Response Team, NASA/GSFC

67   Jacques Descloitres, MODIS Land Rapid Response Team, NASA/GSFC

68   NASA/GSFC, data from NOAA GOES

70   USGS/EROS Data Center

71   Lawrence Ong, EO-1 Mission Science Office, NASA/GSFC

72   NASA/GSFC MODIS Snow & Ice Team

73   Jacques Descloitres, MODIS Land Rapid Response Team, NASA/GSFC

74   Jacques Descloitres, MODIS Land Rapid Response Team, NASA/GSFC

75   SeaWiFS Project, NASA/GSFC and GeoEye

76   South Georgia Island, Jacques Descloitres, MODIS Land Rapid Response Team, NASA/GSFC

76   Canary Islands, NASA/GSFC

76   Attu Island, USGS/EROS Data Center

76   Kuril Islands, USGS/EROS Data Center

77   Kuril Islands, Jacques Descloitres, MODIS Land Rapid Response Team, NASA/GSFC

77   Sado Island, NASA/JSC/Earth Sciences and Image Analysis, photo STS066-117-034

78   National Air and Space Museum

79   National Air and Space Museum

80   GeoEye

81   Aitutaki, Space Imaging

81   Baker Island, Space Imaging

82   Earth Satellite Corporation

83   NASA/GSFC/LaRC/JPL, MISR Team

84   France, Jeff Schmaltz, MODIS Rapid Response Team, NASA/GSFC

84   Great Barrier Reef, NASA/JSC/Earth Sciences and Image Analysis, photo STS009-35-1622

85   Jeff Schmaltz, MODIS Rapid Response Team, NASA/GSFC

86   National Air and Space Museum

88   National Air and Space Museum

90   NASA/GSFC

91   NASA/GSFC

92   Frederick C. Engle, National Air and Space Museum

94   National Air and Space Museum

95   National Air and Space Museum

96   Jeff Schmaltz, MODIS Rapid Response Team, NASA/GSFC

97   Kamchatka Topography, SRTM Team, NASA/JPL/NGA

97   Erupting Volcano, National Air and Space Museum, astronaut photo STS68-150-47

98   GeoEye

99   National Air and Space Museum

100   CNES/SPOT Image

101   NASA/JSC/Earth Sciences and Image Analysis, photo ISS005-E-19017

102   National Air and Space Museum

103   Miyakejima, DigitalGlobe

103   Plume, NASA/GSFC

104   National Air and Space Museum

105   Goma, ImageSat International

105   Virunga Mountains, National Air and Space Museum/NASA/GSFC/METI/ERSDAC/JAROS, NASA/GSFC Landsat 7 Science Team

106   USGS/EROS Data Center

108   GeoEye

109   GeoEye

110   NASA/JSC/Earth Sciences and Image Analysis, photo STS066-208-025

111   NASA/JSC/Earth Sciences and Image Analysis, photo ISS001-E-6765

112   DigitalGlobe

113   National Air and Space Museum

114   National Air and Space Museum

116   National Air and Space Museum

117   USGS/EROS Data Center

118   Satellite Imaging Corporation

119 USGS/EROS Data Center
120 USGS/EROS Data Center
121 Kara-Kul, USGS/EROS Data Center
121 Shoemaker, USGS/EROS Data Center
122 USGS/EROS Data Center
124 USGS/EROS Data Center
126 NASA/GSFC
127 Jennifer Bohlander, NSIDC
130 National Air and Space Museum
131 R&D Center ScanEx
132 NASA/JSC/Earth Sciences and Image Analysis, photo STS026-40-55
133 Mouth of the Amazon, NASA/GSFC/LaRC/JPL, MISR Team
133 Manaus, NASA/GSFC/LaRC/JPL, MISR Team
134 William Stefanov, Arizona State University, NASA/ERSDAC
135 DigitalGlobe
136 National Air and Space Museum
137 NASA/GSFC/METI/ERSDAC/JAROS, U.S./Japan ASTER Science Team
138 NASA/GSFC/METI/ERSDAC/JAROS, U.S./Japan ASTER Science Team
140 Three Gorges, NASA/GSFC/METI/ERSDAC/JAROS, U.S./Japan ASTER Science Team
140 Itaipu Dam, Space Imaging
141 Three Gorges Dam, DigitalGlobe
141 Three Gorges Construction, DigitalGlobe
142 Earth Satellite Corporation
143 USGS/EROS Data Center
144 DigitalGlobe
146 Mississippi, NASA/GSFC/METI/ERSDAC/JAROS, U.S./Japan ASTER Science Team
146 Mekong, NASA/JSC/Earth Sciences and Image Analysis, photo STS075-721-047
147 NASA/JSC/Earth Sciences and Image Analysis, photo STS075-706-082
148 USGS/EROS Data Center
149 Jacques Descloitres, MODIS Land Rapid Response Team, NASA/GSFC
150 USGS/EROS Data Center
151 Jeffrey Kargel, USGS, NASA/JPL
152 Jacques Descloitres, MODIS Land Rapid Response Team, NASA/GSFC
153 USGS/EROS Data Center
154 NASA/GSFC/METI/ERSDAC/JAROS, U.S./Japan ASTER Science Team
155 DigitalGlobe
156–57 NSIDC, NASA/JPL, Canadian Space Agency, Ohio State University, Alaska SAR Facility
158 SeaWiFS Project, NASA/GSFC and GeoEye
159 NASA/GSFC/LaRC/JPL, MISR Team
160 DigitalGlobe
162 ImageSat International
164 NASA/GSFC
165 NASA/GSFC
168 Mauna Kea, GeoEye
168 Mauna Kea, Space Imaging
169 Launch Pad, DigitalGlobe
169 VAB, Space Imaging
170–171 NASA/GSFC/Visualization Analysis Laboratory
172 Mecca, DigitalGlobe
172 Vatican, GeoEye
173 GeoEye
174 DigitalGlobe
175 Great Wall of China, NASA/GSFC/METI/ERSDAC/JAROS, U.S./Japan ASTER Science Team
175 Angor Wat, GeoEye
176 National Air and Space Museum
177 NASA/GSFC/METI/ERSDAC/JAROS, U.S./Japan ASTER Science Team
173 Statue of Liberty, Space Imaging

178 Australia, GeoEye/Satellite Imaging Corporation
179 GeoEye
180 Libya, ImageSat International
180 Kazakhstan, NASA/JSC/Earth Sciences and Image Analysis, photo STS112-E-6013
181 NASA/JSC/Earth Sciences and Image Analysis, photo ISS005-E-21126
182 USGS/EROS Data Center
184 Spain, European Space Agency
184 Arabian Sea, NASA/JPL
185 ImageSat International
186 Jacques Descloitres, MODIS Land Rapid Response Team, NASA/GSFC
187 NASA/JSC/Earth Sciences and Image Analysis, photo STS092-713-32
188 Brasilia, DigitalGlobe
188 Mexico City, DigitalGlobe
188 Rome, DigitalGlobe
188 San Diego, DigitalGlobe
188 Yokahoma, DigitalGlobe
188 Sydney, GeoEye
189 Montreal, DigitalGlobe
189 Abu Dhabi, ImageSat International
189 Pyongyang, ImageSat International
189 Indianapolis, GeoEye
189 Denver, GeoEye
189 Black Rock Desert, GeoEye
190 USGS/NGA
191 GeoEye
192 William Stefanov, Arizona State University, NASA/ERSDAC
193 Las Vegas at Night, William Stefanov, Arizona State University, NASA/ERSDAC
193 São Paulo, William Stefanov, Arizona State University, NASA/ERSDAC
194 GeoEye
195 GeoEye
196 National Air and Space Museum
198 GeoEye
200 NASA/GSFC TOMS Science Team
201 NASA/JPL
202 USGS/EROS Data Center, National Air and Space Museum
204 Boston, DigitalGlobe
204 Detroit, USGS/EROS Data Center
204 Philadelphia, USGS/EROS Data Center, National Air and Space Museum
205 ImageSat International
206 CNES/SPOT Image
207 Robert Simmon, NASA/GSFC Landsat 7 Science Team
208 SRTM Team, NASA/JPL/NGA
209 Panama Canal, National Air and Space Museum
209 Suez Canal, NASA/GSFC/METI/ERSDAC/JAROS, U.S./Japan ASTER Science Team
210 Jacques Descloitres, MODIS Land Rapid Response Team, NASA/GSFC
211 William Stefanov, Arizona State University, NASA/ERSDAC, National Air and Space Museum
212 GeoEye
213 R&D Center ScanEx
214 European Space Agency
215 Gibraltar, NASA/GSFC/METI/ERSDAC/JAROS, U.S./Japan ASTER Science Team
215 Lisbon, GeoEye
216 GeoEye
217 GeoEye
218 National Air and Space Museum
220 National Air and Space Museum
221 ImageSat International
222 DigitalGlobe
223 Atlantic Coast, Jacques Descloitres, MODIS Land Rapid Response Team, NASA/GSFC
223 Cape Town, GeoEye

225 Jacques Descloitres, MODIS Land Rapid Response Team, NASA/GSFC
226 GeoEye
227 GeoEye
228 National Air and Space Museum
229 DigitalGlobe
230 GeoEye
231 JFK Airport, ImageSat International
231 Tokyo, GeoEye
231 Hong Kong, GeoEye
231 Frankfurt, GeoEye
231 Honolulu, ImageSat International
232 San Francisco, CNES/SPOT Image
232 Tokyo, NASA/GSFC/METI/ERSDAC/JAROS, U.S./Japan ASTER Science Team
233 GeoEye
234 USGS/EROS Data Center
236 Antarctica, Jacques Descloitres, MODIS Land Rapid Response Team, NASA/GSFC
236 Global, NOAA/NESDIS
237 USGS/EROS Data Center
238 Fires, L. Giglio & J. Kendall, SSAI, data from TRMM VIRS
238 Landcover, University of Maryland, College Park
240 R&D Center ScanEx
241 Jacques Descloitres, MODIS Land Rapid Response Team, NASA/GSFC
242 NASA/JSC/Earth Sciences and Image Analysis, photo AS07-8-1932
243 USGS/EROS Data Center
244 National Air and Space Museum
245 USGS/EROS Data Center
246 DigitalGlobe
247 GeoEye/CRISP, National University of Singapore
248 NASA/GSFC/LaRC/JPL, MISR Team
249 Jacques Descloitres, MODIS Land Rapid Response Team, NASA/GSFC
250 Jacques Descloitres, MODIS Land Rapid Response Team, NASA/GSFC
251 USGS/EROS Data Center, National Air and Space Museum
252 Jeff Schmaltz, MODIS Rapid Response Team, NASA/GSFC
253 NASA/JSC/Earth Sciences and Image Analysis, photo STS026-43-98
254 Jacques Descloitres, MODIS Land Rapid Response Team, NASA/GSFC
256 National Air and Space Museum
257 Riyadh 2000, William Stefanov, Arizona State University, NASA/ERSDAC
257 Riyadh 1972, National Air and Space Museum
257 Dallas 2001, National Air and Space Museum
258 National Air and Space Museum
259 National Air and Space Museum
260 Jesse Allen, NASA Earth Observatory, USGS/EROS Data Center
261 DigitalGlobe
262 Jacques Descloitres, MODIS Land Rapid Response Team, NASA/GSFC
263 Borneo, Jacques Descloitres, MODIS Land Rapid Response Team, NASA/GSFC
263 Sydney, NASA/GSFC

Index